AF838263

Movement and Emptiness

MOVEMENT AND EMPTINESS

By L C BECKETT

"That moves and That moves not;
That is far and the same is near;
That is within all this and That is
also outside all this".

ISHA UPANISHAD V5

LONDON
STUART & WATKINS

FIRST PUBLISHED IN 1968
BY VINCENT STUART AND JOHN M WATKINS LTD
45 LOWER BELGRAVE STREET
LONDON SW1

PRINTED AND BOUND IN GREAT BRITAIN
BY WORDENS OF CORNWALL LTD
MARAZION, CORNWALL

SBN 7224 0100 0

CONTENTS

ACKNOWLEDGEMENTS

I wish to express my sincere thanks to the following for permission to quote from various works: Allen & Unwin Ltd., Peierls, *Laws of Nature*; Eddington, *Science and the Unseen World*: Cambridge University Press, Eddington, *Nature of the Physical World*; Sherrington, *Man on his Nature*; Schroedinger, *What is Life?*, *Mind and Matter*: The Clarendon Press, Young, *Doubt and Certainty in Science*; Eddington, *Stars and Atoms*: Oxford University Press, Oppenheimer, J. R., *Science and Common Understanding*; Bondi, Lyttleton et alia, *Rival Theories in Cosmology*: The Hutchinson Publishing Group (Rider), Lama Govinda, *Foundations of Tibetan Mysticism*: Victor Gollancz Ltd., Barnett, *The Universe and Dr. Einstein*; Pfeiffer, J., *The Changing Universe*.

To some this book may appear to be an intellectual attempt to rationalise that which has always been impossible to describe. But as I feel myself to be a child of my time, this 20th century, which has an essentially scientific turn of mind, I have taken the way of my century to approach ancient truths. In taking this path however, I have discovered that even in our day, starting from the periphery one may work inwards to the core, without even seeking such a result. Without an aim I might say, by letting my thoughts wander as they would in the worlds of galaxies and atoms, I have slowly realised that I was being led back to the one fundamental fact of the "Unborn, Unoriginated, Unformed" which the Buddha recognised. In science as in religion it is inescapable. What had seemed to be a purely scientific conception proved to be but the external aspect of what lies in the depths—in "nothing".

Whoever thinks at the end of it: this is 'thinking' not 'seeing' has not understood.

LCB

I

MOVEMENT

"The atom of action seems to have no coherence in space; it has a unity which overleaps space".

EDDINGTON

THROUGHOUT THE HISTORY OF THE HUMAN race, as far back as we can trace it, we find that man has been engaged in trying to answer the question: what is Reality? And instead of accepting the world around him as 'real', which would have been logical, he has insisted on seeing the signs of Reality in the things he did not understand, which did not coincide with his experience of life—strange as this may seem. In these he has seen manifestations of Powers that rule life on Earth. Today as before, we are looking to Sun, planets, stars—or atomic structures, for a solution to our problem. What is all this in Reality? How did it arise? Who or what created it?

It is strange if one comes to think of it that the answer should not have been discovered long since in electricity or magnetism—if the physical world were not answer enough. But those magic appearances were ignored and instead the answer was sought after in the image-ination. However much the ideas of God (or Gods) differed according to time and place where they arose, for every race, for the most primitive as for the most evolved, the answer to this question of 'Reality' was always found outside this phenomenal world.

Against the bar of that Reality, men's lives have been (and are) judged: whether it be God in the sky or God in the snake totem, IT rules men's lives. Hence we can only conclude that this idea must arise in an awareness of some unknown factor ingrained in the mind of man. Modern psychology would call it an 'archetype'. But what is a tremendous thought: such idea has clearly evolved together with the evolution of *homo sapiens*: from Gods that were as much of this world as were men and animals, to the God who is pure spirit, men's Gods have grown in proportion to the creative awareness of their minds. But at all times and places, these images have represented TRUTH—something inherent in, and at the same time beyond, our lives on Earth.

To-day we seem to have no generally accepted Truth even in any one community; we are all looking in every direction for a catalyst that shall solve our troubles, and though many are still devout Christians, Buddhists or Mohammedans, all equally are assimilating beliefs which are rooted in a tradition totally different from their own, which bears little relation to the present. What connection is there between our restless, tortured, mass-controlled existence, facing threats from every side, and the peaceful agricultural communities of Magadha or Palestine? But a belief that is to help men live their lives should correspond to the circumstances which have given them rise. Religions so far answered a need—where can we find an answer to our need to-day? It seems to me that we find ourselves in the same chaos, religiously speaking, as that which prevailed in the Roman world of the first century of our era. Man had then outgrown the old beliefs, these no longer sufficed in an age which had known some of the greatest thinkers of all-time, so men turned to a new teaching which seemed to help them live their lives a little more serenely. Can we say that any of the old beliefs do that for us? So long as things go well we can be most devout, but when bombs rain down from the skies, or our lives are fraught with alarms and difficulties, do we appeal to religion for help? Even time is too short.

I think that the mistake we make lies in our insistence in regarding religion as contained in the forms handed down by our ancestors, whilst at the same time our creative awareness stretches out to the limits of the Universe. If 'religion' is to keep pace with the progress of our minds, (as it has in the past) we shall have to re-examine our point of view altogether. Accepting the probability that every religion has contained some aspect of Truth, "seen as through a glass darkly", (why otherwise have one and all exerted so great an influence on the evolution of man?) should we not ask ourselves whether in our time there may not be a fresh manifestation of such truth; unrecognised by us, because couched in terms we have never regarded as bearing any relation to religion but exerting as great an influence as ever a religion has. If that be so, it is high time we examined this matter very carefully in order to find out wherein lies *our* Truth.

Today all races are uniting in an effort to probe the secrets of nature: the world of science belongs as much to the American as to the Russian or ourselves, in this matter there are no frontiers between nations, all are one family in the quest of knowledge. If Sir Bernard Lovell wants information on flare stars as sources of Radio waves, he collates the findings of Radio astronomers at Shiraz in Persia, Olifontein in South Africa, San Fernando in Spain, Curaçao in the Dutch Antilles and Villa Dolores in the Argentine, with what he has observed at Jodrell Bank. Is it not possible that such collective efforts of mankind may bring about a new understanding of that Unknown which has heretofore eluded us, and forge links in a chain of universal unity—a goal that has proved unattainable until now? As the greatest of our scientists remarked: "The pursuit of science springs from a striving which the mind is impelled to follow, a questioning which will not be suppressed. (Was that not the origin of every religion? LCB) Whether in the intellectual pursuits of science or in the mystical pursuits of the spirit, the light beckons ahead and the purpose surging in our nature responds".[1] That light has illuminated all faiths—that purpose has created all Gods. As Meister Eckhart said: "That God is 'God', of that I am the cause",[2]—equally that Siva is "Siva", or Allah "Allah". At all times and places the "I" has created images to which it could ascribe its fate, or turn in the hour of need.

The question which we may ask ourselves today is: were I not, would stars still be stars, atoms still be atoms? Were there no human eye to behold the stars, would they still lie in the heavens? This faces us with the problem of the "reality" or "illusion" of both Gods and stars, the old question that has always tormented

[1] Sir Arthur Eddington: *Nature of the Physical World*, p.328.
[2] Eckhart: Buttner transcription, vol. I, p.198.

mankind. Even admitting that our present ideas may turn out to be as fantastic as were all the rest, yet they are born from out of the soil we are cultivating in our time, and should therefore be more of a reality to us than are those we have inherited from the past. Is it not possible that we have reached a stage in development where an awareness arising out of human consciousness, from out of a perception of "what is", is no more the prerogative of one race, but is the property of all mankind? Even that is not necessarily the limit: some day we may perhaps share our knowledge with other stars; but every step on this path may bring us that much nearer the Truth.

The most pertinent question is: what do we regard as "Reality" today? May the answer perhaps lie in what we call Science? This suggestion is not so far-fetched as would appear, for in its time every religion has had a close connection with what is now regarded as the domain of science. The Cosmos has always been the home of the Gods: in the Sun or behind the clouds *they* were always "up there". But if this science of ours, with which all men are equally concerned, is to take the place of religion, then it behoves us to try and understand what it really means, not merely as intellectual theory, but as a reality which affects our lives. Could there possibly be more "reality" in this than in all imagery of the past? Or are all equally attempts at climbing a mountain from different angles? But whatever our starting point today, what we believe in as fact, must hold as much of reality for us as did the Gods for earlier man, therefore we should try and understand what truth there may be *for us* in certain occurrences about which we make definite statements and to which we apply names hitherto unknown. Can we grasp the true meaning of the words we use when we say: "a billion light-years" separate us from some star? Or when we measure the size of an atom by "a millionth of an inch"? Even the composition of our own bodies has proved to be quite different from what we had imagined. Every day scientists discover new facts about the world, but are they really so new? Are we not perhaps applying new terms,—concordant with our present-day outlook—to that which was regarded as "Gods" in ancient times, and seeing that which has always lain at the back of human consciousness from a new perspective? Reality underlies many coats of paint, and we gaze at the spectacle before us as we gaze at abstract pictures—putting our own interpretation. But that does not exempt us from trying to find out what lies behind this coat of many names. I believe that somewhere within us lies a knowledge of what is *real* and what is not, in fact I suspect that underneath all the symbols that we have attached to religious ideas, there is something that only a few enlightened seers have contemplated—scientists as well as mystics,—and which they have attempted to formulate, each in his own terms. "You should not start reasoning from perceptions, (seeing, hearing, feeling, knowing) nor allow your thinking to stem from them . . . Neither hold to them, abandon them, dwell in them nor cling to them, but start independently of all that is above, below and around you, for there is nowhere in which the Way cannot be followed."[3]

[3] Huang-po: *Doctrine of Universal Mind*, Section 9.

Perhaps the first step on this new way might be a revaluation of some of the facts to which science is opening our minds. We might possibly discover in the world of stars and atoms a new upsurging of that unknown factor in the collective unconscious—this time of all mankind—which has been the source of the many religious beliefs in the past. As beginning we might consider what has happened to all our ideas of *mass*, that rock on which we built our lives. Science has opened our eyes to the fact that the solidity of this table which seems to me an undeniable fact, is not any more like what I think it is than it is like what the physicist describes—it is neither the one nor the other, yet both. In fact we have no certainty that anything is 'in reality' what our minds have constructed: the Principle of Uncertainty has seen to that. The same of course applies to the Gods. "All names which the soul gives to God, those she draws from herself" as Meister Eckhart truly remarked.[4] She draws the *names* but not what lies behind them. Why should we imagine that the soul, or what C. G. Jung called the "Unconscious", could not draw new names from out of itself for that Unknown of which it alone has some perception thus making us see it with new eyes? Why should we reject new terms merely because they have so far represented physical entities to our minds? If there be something today which is as significant to us as were the Gods in ancient times, it must be manifesting in some form we are not yet aware of, under some name we do not recognise. But if we aspire at seeing things

[4] Franz Pfeiffer transcription translated from the German by the author.

that were held sacred by our ancestors from a new point of view we should make a start by looking at them in terms that are new to us. This may require a complete metanoia (turn about), but this every religion has demanded.

Mystics at all times and places have sought to find the reality underlying this apparent life, in something within us, call it Mind or Soul. Today mystics, humanists and scientists discover in the human mind seeds of future evolution, but the trouble is that our bodies are not keeping pace with our minds, they have remained unchanged for thousands of years, while the mind goes soaring into the depths of space and time. And unless we can find some means of balancing the two and lighten the mind of this troublesome burden, we shall be torn asunder between mind and matter *in ourselves*. We shall evidently have to find some way of doing as Eckhart suggests: "When the spirit turns away from all things that have become (been created LCB) into its very self in the unbecome of the eternal image . . . this means the spirit turns back into its origins".[5] So far man has tried to find "reality" in the "become"—what may he not discover if he stumbles into the "unbecome"? But I should warn my reader that the further we proceed on this road, the more shall we find ourselves retreating from human dimensions. The aim of every religion has been to pierce the veil of what our senses apprehend, (being part of it) but we have only been able to make more or less valid guesses at that which is unapprehensible to them. All I hope is that I may be able to formulate this question in terms

[5] F. Pfeiffer transcription, p.517.

familiar to us at the present time, whilst at the same time showing it not limited by those terms or any other. An understanding of my objective depends entirely on the approach of the reader, for it has nothing to do with material concepts, even though couched in terms of them. In the final instance it is nearer to no-thing.

II

WHATEVER WE MAY THINK ABOUT OUR situation in the universe, there can be no doubt but that we are denizens of the solar system, whatever that may be in "reality". And furthermore, as Fred Hoyle says: "we cannot think ourselves outside the particular patterns that our brains are conditioned to, or, to be more accurate, we can think only a little way outside them, and then only if we are very original".[1] One of the most "original" and pregnant ways of thought in the history of the human mind seems to me to be contained in the Lankavatara Sutra, and although it is about 2,000 years old, in our time we seem to be rediscovering its message.

Intrinsic to the Lankavatara Sutra is the idea that everything in the universe is nought but Mind-only, and seeing that we, today, are no more able to regard matter as solid mass, as which it was always regarded heretofore, we may at least take as starting point for our view of the world, the possibility that if it is not material it may be mental and reconsider the teaching of the Lankavatara. But before we begin, it must be understood that the premiss for all the ideas

[1] *On Men and Galaxies*, p.59.

of the Lankavatara Sutra is the assumption that the whole world, *as we see it*, is an illusion created by the five senses ("skandhas" in Buddhist terms); I do not think that any student of physics today would quarrel with that assumption however, certainly neither Eddington nor Sherrington, according to what I know of their ideas.

As there is one chapter of this Sutra which has inspired all the ideas contained in the following pages, I should like to give an extract from it. It is said:

> "When it is understood that there is nothing in the world but what is seen of the Mind itself, discrimination no more arises . . . The ignorant work and discriminate, but not the wise. Mahamati, the doings of the ignorant are unrealities discriminated, realities confounded; they are like the City of the Gandharvas, like magically-created figures. To illustrate, Mahamati, here is a City of the Gandharvas where children see magically created people, merchants and many others, going in or coming out".[2]

Eddington after a life devoted to astronomy and physics, arrived at the same conclusion: "When the environment of space and time and matter, of light and colour and concrete things . . . is probed deeply by every device of physical science, at the bottom we reach symbols. (The 'City of the Gandharvas' LCB). Its substance has melted into shadow".[3]

But the Lankavatara Sutra adds something more which I think holds the key to things which are now being discovered through science: "It is owing to the discrimination characterised by perturbation that such (the magic creation) takes place". In other words: it is motion that causes us to discriminate—without

[2] Ch. LXXVII. [3] *Science and the Unseen World*, p.24.

movement (perturbation) we should not apprehend anything that we could use as material for our magic creation. But, at the same time we must realise that we could not "discriminate" perturbation were there not something unperturbed. Equally were there no movement we could not discriminate quiescence—what many people call the "void". It is clear therefore that these two are at the foundation of our image of the City of the Gandharvas. So the premiss to any approach to "reality", whatever that may be, seems to lie in a dual factor which impresses itself on our consciousness.

This idea of the universe as a duality was first conceived of by the Chinese several thousand years ago. They classified the whole world as Yang and Yin, which encompassed all dualities: light and shade, positive and negative, plus and minus, male and female, creative energy and death. There are even Yang and Yin people, Yang and Yin foods. According to the Chinese teaching one should start by realising one is either of Yang or Yin nature, and order one's life accordingly. Basic to Buddhism is the idea of Samsara and Nirvana; and I think Einstein understood this same fact in modern terms as "space-time", which pushes duality out to the limits of the universe. He died still trying to discover the unification of that duality. I could quite legitimately base myself on one of these three expressions, but as I do not give the orthodox interpretation to either one of them, I prefer to use terms of my own. Space is commonly regarded as the three-dimensional condition of the physical world, the world as revealed by our five senses, plus brain;

and time is the dimension in which all occurs. But to my understanding "space" is empty space, void, a condition of no-thing, what Zen Buddhists call: Sunyata, (emptiness) taking that to mean that: "Emptiness is beyond eternalism and nihilism", as the Lankavatara says. "Time" to my mind means movement, movement *as such*, not movement related to any object *in* motion. In this context objects do not exist. Therefore I propose to call the great duality: Movement and Emptiness.

In this I am not running counter to scientific tenets, for in the course of this century scientists have come to realise that there is no solid body at all, but only larger or smaller systems *in motion*. As Einstein expresses it: "Space has no objective reality except as an order or arrangement of the objects we perceive in it, and time has no independent existence apart from the order of events by which we measure it."[4] That applies to both cosmic and atomic systems, all equally are measured by the "order" of their movements, they are all "events" we see in relation to one another, across that which has no "objective reality". It would seem that the only thing we can be sure of is movement. According to present-day cosmologists: "the Universe is a restless place: stars, nebulae, galaxies, and all the vast gravitational systems of outer space, (and I may add: all atomic systems as well LCB) are incessantly in motion. But their movements can be described only with respect to each other (and ourselves LCB) for in space there are no directions and no boundaries".[5] But without the empty space between, we could not recognise move-

[4] Lincoln Barnett: *The Universe and Dr. Einstein*, p.14. [5] ibid, p.39.

ment of any kind, seeing that we can only know things *are*, or what we think they are, by registering their motions relative to one another, and these motions could not take place in a fixed fullness. As a matter of fact that is exactly what the universe looked like until Einstein proved we were mistaken! From the times of the Greeks till the 1930's everything was believed to be floating in a universal ether; it was considered impossible that heavenly bodies should float in nothing. Consequently everything we saw in the heavens, as well as this Earth itself, and all its contents subsisted in an all-pervasive jelly-like substance in which bits of banana or orange or cherry were encased. But in this last half century we have had to recognise that anything in motion can travel by its own impetus and needs no jelly to support it. We have proof of this in the vacuum tubes in our laboratories, the best of which contain about 100 million particles, all moving quite happily. So Einstein abolished the idea of ether, the existence of which was after all, only based on induction.

W. H. Bonner gives a graphic description of what I am trying to explain and what he says is the best introduction I can find. Accepting the fact that Einstein's discovery of the General Theory of Relativity completely altered the ideas of space-time, he says: "the situation now is one of compromise between the void and the plenum. On the one hand ... matter is atomic: there certainly exist discrete particles separated by space. On the other hand, a piece of matter has a sphere of influence called its field; one may think of the gravitational field of earth or the electric field of a proton. In so far as a field means a region of influence *there is nothing in it which contradicts the idea of empty space* . . . What we usually think of as empty space contains the energy of numerous fields—gravitational, electro magnetic and the fields of the quantum theory. Embedded in the fields are the particles of matter which are to be regarded as their sources. They are continuous quantities spreading through otherwise empty space".[6] Quantities of movement as we shall see!

As the question of what is moved or moving is fundamental to my conception of reality, I think I should say a few words about the laws of motion.

As accepted so far, they dated back to Galileo and Newton and concerned all bodies in the universe, both great and small. Until now, their theories were undisputed. Galileo proclaimed that: "unless a body is acted on by a force, it will continue to move in a straight line with constant speed". The difficulty lay in deciding what determines that constant velocity called the "inertial frame"? As an explanation, Newton's Second Law of Motion was substituted for Galileo's. That would mean that the rate of change and momentum of a body is equal to the external force acting on the body. But ideas on the subject are changing and Hoyle comments: "Plainly the body will not be acted on by a force if it exists by itself in an otherwise empty universe. It is trivial to the mathematician to show that there is not just one unique way of describing the motion of the body, that there are infinitely more ways of doing so, and that in an otherwise empty universe all ways have equal status".

[6] Einstein and the Nature of Space. *Listener* 26/3/59. (italics mine).

Obviously the momentum would remain constant if the external force is zero. The problem remained until Ernest Mach, some 80 years ago, suggested a solution. His idea was: "Suppose there is no mass at all, then the deduction of the first law from the second collapses, in fact the theory itself collapses." From this Hoyle concludes that: "If you follow Mach in this way, you inevitably reach the conclusion that the attribute of mass, which has been present in physics since the time of Galileo *is not an inherent property of matter*, it arises from the interaction of matter with the rest of the universe—take away the universe and the property of mass simply disappears."[7] That conclusion leaves us with nothing but motion itself!

[7] A New Kind of Physics. *Listener* 18/6/64.

III

In all that follows, I am going to make an attempt at explaining my ideas of movement and emptiness as clearly as possible, difficult though that is, but before I begin I must stress one point which is perhaps a commonplace, yet it needs bearing in mind, namely that our knowledge of everything is conditioned by the duality "is" and "is not". The Lankavatara expresses it very clearly: "Again Mahàmati, what is meant by emptiness and mutual non-existence? It is this: when a thing is missing here one speaks of its being there. For instance Mahàmati, in the lecture hall of the Mrigarama there are no elephants, no bulls, no sheep, but as to the Bhikshus I can say that the hall is not devoid of them; it is only empty as far as they (the animals) are concerned. Further Mahàmati, it is not that the lecture hall is devoid of its own characteristics, nor that the Bhikshu is devoid of this Bhikshuhood, nor that in some places too, elephants, bulls or sheep are not to be found. Mahàmati, here one sees all things in their aspect of individuality and generality, but from the point of view of mutuality (itaratara) some things do not exist somewhere. Thus one speaks of the emptiness of mutual non-existence".[1]

In order to discover the truth of this matter, I shall have to begin by setting aside all our traditional beliefs, and start afresh with what astronomers and physicists are discovering today regarding stars and atoms. Beginning with the widest conditions we know of, I shall end with the smallest; we shall see where that may lead.

As Fred Hoyle is my authority for most of what I know about cosmology, I feel I must start my investigations by giving some examples of his ideas. He says: "Just as the basic step in explaining the origin of stars is the recognition that a tenuous gas pervades the space within a galaxy, so the basic step in explaining the origin of the galaxies is the recognition that a still more tenuous gas fills *the whole of space*".[2] It is important therefore to understand what this gas consists of, and fortunately cosmic rays provide the answer. From those we learn that approximately 93 per cent of the total number of atoms in the universe are hydrogen atoms, 7 per cent helium and little more than 1 per cent of all other elements put together. All the various theories about the origins of the universe, however much they may differ in other respects, are agreed that it did not start with ready-made galaxies or stars, but what may be looked upon as the foundation of all the rest is "a cold dilute, and turbulent gas of hydrogen atoms",[3] or perhaps

[1] Ch. XXVII.

[2] *Nature of the Universe*, p.95, *Hoyle*.

[3] William Fowler: The Origin of the Elements. *Scientific American*, September, 1956.

a lump of them, if we believe in the primeval atom? From there, the next step is interstellar dust, the size of the particles of which suggest that it was formed from out of interstellar gas. Like the sun's atmosphere, it is again mostly hydrogen, then helium, some carbon, nitrogen, oxygen and a few metals, all and sundry having the hydrogen atom as foundation. Hubble reached the conclusion that in the universe as a whole there is 000,000,000,000,000,000,000,000,000,001 grammes of matter per cubic centimeter of space. Hoyle remarks: "Small as this is, the total amount of the background material exceeds about a thousandfold the combined quantity of material in all the (million million LCB) galaxies put together". (*ibid*). For me the important point in all this is that, as Herman Bondi remarked, in any attempt to form a theory of the universe, we always come up with the answer "that it is *almost bound to be in motion* with objects showing velocities proportional to their distance . . . If we divide the distance of any galaxy by its velocity of recession, we get the same number whatever galaxy we choose . . . This number is a time (in my terms motion LCB) a time which according to the most recent work is about 10,000 million years. In some way or other, this is the characteristic time of the universe".[4] I think this may enable us to imagine something of the magnitude of the conditions we are facing.

How does the knowledge of these things reach us? We have of course all heard of the waves of light spreading throughout the universe; the distance of its sources are calculated by means of the Red Shift of

[4] Hermann Bondi, *Illustrated London News*, 31/10/59.

the light that reaches our telescopes, but I cannot go into this here. In the last few years however, something much more exciting has happened: a source of information has been discovered that carries us into hitherto undreamt-of depths of the universe—namely Radio waves. John Pfeiffer says: "with them we are in process of building a different kind of synthetic sense organ. Microscopes, conventional telescopes, microphones—such instruments extend the senses we already have. But Radio waves are an example of devices which in effect endow us with new senses. As far as Radio waves are concerned, we start almost from scratch, as if we were born blind and were suddenly to see for the first time".[5] Radio waves arrive here from the Sun, stars and even from the Moon. They are a million times longer than light waves; the brightest of all Radio stars "turns out to be something never before observed in the history of astronomy". Named Cass from Cassiopia in whose direction lies its source, "it broadcasts at the amazing power-level of 10 million billion billion kilowats. Looking at Cass . . . it is impossible not to think of waves and surf on a galactic scale. The twisting filaments seem to be concentrated round a particular region, as if something there in the centre—perhaps a vast invisible whirlpool or magnetic field—were stirring things up. It is almost as if you were looking at an enormous lake in the sky. You do not see the unruffled surface but only the rough parts, white caps of a sort, whipped up by cosmic winds",[6] only of course we are not "seeing" with our eyes nor "hearing" with our ears;

[5] *The Changing Universe*, p.35. [6] ibid, p.97.

our knowledge would seem to stem from a new sense, possessing qualities of both. Lately new sources of these waves have been discovered called "quasars" at distances so fantastic we can hardly grasp them. All this can be summed up in Pfeiffer's words: "Matter in motion is universal and so is radiation. The two are always associated. A universe without radiation would be cold and dead".

But it is not only through Radio telescopes that we have learnt so much; before the advent of these, as our telescopes became more and more powerful, the spectrum of stars was charted and from this the measurement of the rotation of stars was made possible. In the course of intensive study of the stars, especially since the great telescope at Mount Palomar and the Radio telescopes were built, it has been found that all stars are rotating, and that: "nearly every cluster has its own characteristic association of rotational velocity with mass". For instance the heaviest stars in the Pleiades rotate faster than the field stars of the same mass, but the stars of the Pleiades of less than four times the mass of the sun, rotate unusually slowly. On the other hand, the most massive stars in the Hyades and Praesepe clusters, rotate rather slowly and the lighter ones rotate more rapidly".[7] It looks as if the speed of the rotation and mass are somehow correlated; only binary stars and those having planets seem to have different systems of rotation.

But of the fact that all stars and galaxies, massive or not, rotate as well as moving with the expansion of the universe, there can be no doubt.

[7] Helmut Abt: Rotation of Stars, *Scientific American*, February, 1963.

IV

THAT BRINGS ME TO THE NEXT IMPORTANT question that is facing science today, and which concerns all I have been saying very deeply, namely *mass*. What is that, what does it consist of? We saw above that 93 per cent of the whole material universe is composed of hydrogen atoms; in fact it is said that: "All galaxies, stars, planets, and human beings are manifestations of disturbances in a tenuous sea. They are waves, streams, ripples, flurries in ceaseless change. Only that sea of cosmic gas remains"[1]—that gas of what we call hydrogen. Being the root atom of the whole universe, if we want to learn about what is fundamental to creation itself, we shall have to consider this atom very closely. Pfeiffer says that: "to observe the process of creation we must observe hydrogen",[2] and I do not think even the present concentration on quasars will alter that fact. But I should make it clear that whatever I have discovered of its nature was not the fruit of any preconceived idea on my part. It was Fred Hoyle who first drew my attention to the importance of hydrogen in the universe,[3] and what he said was confirmed by

[1] J. Pfeiffer opus cit., p.154. [2] ibid, p.156. [3] See *Nature of the Universe*.

R. A. Lyttleton, Herman Bondi, Peierls and others. In fact as I proceeded I found that there is no book or article on physics that does not take this into account.

Most people today know something about the atomic world—nuclear physics have brought it very close to us—therefore if I go into the question of the structure of the hydrogen atom, I am not touching upon something that is the prerogative of physicists, but I intend to regard it from an even wider angle than theirs, for I hope to make it clear in the end, that it is the prototype of all manifestations in the universe.

First of all we should learn what this all-important atom consists of: each atom is about a millionth of an inch in diameter and weighs 1.6×10^{-24} grams; Eddington exclaims that: "imagination balks before anything so minute"! The hydrogen atom is composed of a nucleus made up of proton and neutron, and an electron. Every atom is the same size and possesses the same properties as every other of the same kind. "They last indefinitely. We think of them rightly as completely stable and unchanging. When they are undisturbed they do not radiate light or any other electro-magnetic radiation. When they are disturbed, they sometimes do radiate but the colour of the light they emit is not random and continuous, but falls in the sharp lines of the hydrogen spectrum".[4] Niels Bohr postulated that: "in every atom there were stationary states whose stability and uniqueness would not be understood in terms of classical dynamics. The lowest one, the one with the least energy, the ground

[4] J. Robert Oppenheimer: *Science and the Common Understanding*, pp.44-5.

state, is truly stable". (As we shall see later, "stable" does not mean motionless but "always the same" LCB). "The others are called excited states and they may be excited by collision or radiation or other disturbance. They too are stable . . . (but) their stability is not absolute. Just as these excited states could be reached by transition, induced by collision or disturbance, so an atom may return to states of lower energy whether by further collision or spontaneously. In these spontaneous changes it gives out radiation which is the analog of that radiation which in classical theory would make all motion unstable".[5] According to these theories the energy of the stationary states gives rise to the idea of the electron moving in orbits round the nucleus—the so-called "quantum" conditions, but Oppenheimer remarks: "We learned to accept as we later learned to understand that the behaviour of the atomic system is not predictable in detail . . . nowhere in our battery of experimental probings could we find what an individual atom would do". (*ibid*). Only statistical prediction is now possible.

What is to my mind the most important lesson to be learnt from all this is that concerning the atomic *transitional* states, and it would seem that this was also what Oppenheimer was leading up to, for in the end he asks: "What are we to think of the transitions themselves? Do they take place suddenly? Are they very quick motions, executed in going from one orbit to another? Are they causally determined? Can we say, that is, when an atom will pass from one of its states to another as we disturb it; and can we find what

[5] Opus cit., pp.48–9.

it is that determines that time? To all these questions the answer would turn out to be 'no' . . . What we needed to understand was not the state of affairs during the transition but the impossibility of visualizing the transition . . . *in terms of the motion of matter*". Which fact he sums up in terms that open up an entirely new aspect of the universe, because such a thing had been inconceivable until now: "We saw in the very heart of the physical world, an end to complete causality",[6] that just where we would have expected to find it most firmly established! This reveals the first and essential quality of the hydrogen atom: it is entirely independent of our intentions or attempts at interference.

Yet we have been told that there is "mass", and for that, logically speaking, there should be some cause. In any effort to try and understand what "mass", (what we have always presumed to be "matter") might be, in "reality", we shall have to make an effort to understand the composition of that hydrogen atom which R. A. Lyttleton describes as "the primitive element in the universe, from which it is believed all the other elements are built up by nuclear synthesis in the stars".[7] To begin with we must realise that all atoms in the whole universe are possessed of the same ingredients as the atom of hydrogen in varying quantities: 1 electron in hydrogen, 4 in carbon, 13 in aluminium etc. up to 92 in uranium. What does an atom look like? To begin with we must expunge from our minds any idea of its being a bullet-like structure; it is on the contrary extremely tenuous,

[6] Opus cit., pp.50. (italics mine). [7] *Illustrated London News*, 18/4/59.

in fact we shall find that it is only recognisable by the velocity of its motions. The atom as a whole is neutral and 10,000 to 100,000 times as far across as its tiny nuclear core. In fact Prof. R. E. Peierls says: "It is about as empty as the solar system, i.e. the atomic nucleus is as small by comparison to the size of the atom as the Sun in comparison to the orbits of the planets".[8] Here we find atom and emptiness—our first glimpse of the dual nature of the Universe.

As the hydrogen atom is electrically neutral, and the electrons were found to be negative, in order to balance that fact, the total charge on the nucleus had to be positive, consequently the first question that arose in this connection was: is the charge distributed in some way throughout the atom or is it concentrated at the centre? From experiments in firing Alpha particles on metal foil, Rutherford deduced that there must be some object as heavy as, or probably heavier than the Alpha particle, inside the atom. . . (He) estimated that the size of (it) was no more than one ten thousandth part of the radius of the atom—that is 10^{-12} centimeters (one millionth of a millionth of a centimeter!). This minute object the proton or nucleus may be regarded "as the sun of the atomic system, the matrix of all forces in the universe". And, not only does it carry the whole positive charge but it contains more than 99.9 per cent of the whole weight of the atom. "During the ten years up to 1939 the structure of the nucleus was worked out. It is like a small sphere and it contains two kinds of particles: protons and neutrons". (Since then so-called "elemental" particles

[8] *The Laws of Nature*, p.97.

have been discovered, but of that later). According to Prof. Peierls, the discovery of the neutron was a very important step in the understanding of the nucleus. "This is a particle approximately the same mass as the proton . . . but distinguished from the proton by carrying no electrical charge, (being electrically neutral LCB). That both, protons and neutrons can be knocked out of nuclei as the result of collisions suggests that they are contained in every nucleus, and in fact we know that nuclei are entirely composed of neutrons and protons",[9] together known as nucleons.

Concerning nucleons Prof. Hamilton says: "Clearly they are very basic objects in nature. So we ask what do they look like? Are they just geometrical points or do they themselves have structure?" At one time they were compared to billiard balls but this did not fit the case for they were discovered to have spin, "they are more like cricket balls in flight". Furthermore: "in the case of the proton which has unit positive charge, this spinning motion should give rise to magnetic effect".[10] As O. R. Frisch explains: "a hydrogen molecule, (composed of two atoms of hydrogen LCB) does not rotate like a rigid body: its two nuclei, the two protons, do swing round like a waltzing couple".[11] As a result of this, the proton was found to have $2\frac{1}{2}$ times as much magnetism as had been supposed. But what is more surprising is that the neutron also has a magnetic effect although its charge is zero. According to Peierls: "when two particles spin on their axes in the same direction as that in which they revolve about each

[9] Opus cit., pp.237-8. [10] Report: "Not so elemental", *Listener* 26/10/61.
[11] The Magnetic Proton, *Listener* 26/10/63.

other, the attraction between them is stronger; when the spin is in the opposite direction as that in which they revolve, the attraction is weaker".[12] Therefore it is clear that *motion* is the determining factor.

The most puzzling thing about the relation of proton to neutron is that *both* have a spinning motion at about 1/5 that of the speed of light, which should preclude their remaining together as nucleon. It would in fact be impossible were there not some strong force holding them together. In trying to discover such a force, especially with regard to the fact that it must necessarily be of very short range and great strength, the Japanese physicist Hideki Yukawa proposed that a nuclear quantum of a mass 2—300 times that of an electron, must be exchanged between proton and neutron, at a very rapid rate. Some years later—in 1947 to be exact—those quanta or particles, the pi-mesons or pions, were discovered with just the properties expected. "According to our present picture the proton makes pions and catches them again all the time, like a juggler".[13] Pions have however raised very serious questions which have not yet been satisfactorily answered. When the nucleons eject pions, according to the Law of Conservation of Mass, the mass of the nucleus should be reduced, yet all experiments have shown that *it is always the same*. The answer, as far as we know it, would seem to be that the emission and reabsorption of pions takes place so rapidly "that we cannot detect it. Since any phenomenon that is undetectable cannot be regarded as 'real' in the physical

[12] The Atomic Nucleus, *Scientific American*, January 1959.
[13] Frisch, opus cit.

sense, we must speak of 'virtual' emission and exchange of pions".[14] With the pion we find ourselves faced with one of the most puzzling facts in physical science, namely the necessity of accepting an Uncertainty Principle. It was found unavoidable seeing that there is a limit to what can be known of any small particle. "The more certain we are about the time at which the energy is determined, the less certain we can be about its amount . . . The calculation based on (Planck's) constant h (fundamental to both the quantum theory and the Uncertainty Principle LCB) tells us that the margin of uncertainty for the energy content of a nucleon will be at least 135 MEV, when the time of measurement is 5×10^{-24} of a second. Hence if a pion is emitted and reabsorbed within this time, it will be undetectable i.e. 'virtual'. This range is just the range of nuclear forces! The agreement seems a striking support of the theory that pions act as agents of these forces". (*ibid*).

In speaking of these components of the hydrogen nucleus I feel I may have given the impression that they are "things", such as those registered by our senses; if so I have given a wrong impression. But I admit that it is difficult to believe that entities which seem to be nothing but "spin" could be vital to the structure of the universe. Keeping in mind the "vast invisible whirlpool" of Cass, however, and applying this image to the atom (and it is certainly conceivable that everything in the Universe follows the same pattern) I think we gain some perception of what that image may look like in every dimension. But essentially

[14] Robert Marshak, PIONS, *Scientific American*, January, 1957.

everything seems to come within the range of that "magically created world" prospected by the Lankavatara Sutra, the reality of which lies in our imageination.

V

WE MUST NOW TURN TO THE QUESTION of the electron, that active but negative partner of the nucleus in the composition of all atoms in the universe. Its role—although in itself it is certainly insubstantial—is ubiquitous, it cannot be pin-pointed, yet nothing seems to happen without it.

In the course of work on the quanta of light rays, (which were regarded as waves) it was found that, at times they acquired the character of particles. Peierls, undoubtedly one of our greatest physicists, explains what is known of this strange fact very clearly: "This principle (of uncertainty) says that there is a limit to the accuracy with which we can study the motion of a small object . . . because in doing so we must of necessity expose the objects to a collision with light quanta (which we use for observing it LCB) which will change its momentum by any amount which may be as large as h/d—where h is Planck's quantum . . . Therefore, even if we had previously known the momentum of the particle accurately, it will now be unknown by any amount of h/d . . . Conversely if we know to start with where the particle is located and make an observation which will tell us its velocity to high accuracy, this must, of necessity, involve such a disturbance of its motion that in the end its position will be uncertain by a corresponding amount".[1] That disturbance was called "electron", but as a matter of fact the electron as "object" can never be fixed by any manner of means, as Peierls himself admits: "The electron can never have a physical reality", because it can never be objectified independently of the means chosen for observing it. Accepting it as wave, (being the alternative to "object") leaves us just as much in the dark, however. As Oppenheimer points out: "The spreading of waves in space did not mean that the electron spread; it meant that the probability or likelihood of our finding the electron when we look for it, spreads as the wave does";[2] and such waves may spread indefinitely!

We discover here for the first time direct evidence that one of the principle factors in the physical world is forever undiscoverable because its *continuous motion* defeats all efforts at finding it. But electrons seem unpredictable from every point of view; they are undoubtedly closely related to the nucleus, yet: "if we ask for instance whether the position of the electron remains the same, we must say 'no'; if we ask whether the electron's position changes with time, we must say 'no'; if we ask if the electron is at rest we must say 'no'; if we ask whether it is in motion, we must say 'no'."[3] Here Oppenheimer joins hands with the mystics of all time in their approach to the Transcendental, in fact he himself sees a resemblance in his replies to the answers the Buddha gave to this

[1] Opus cit., p.169 [2] Opus cit., p.61. [3] Opus cit., p.43.

kind of speculation. Yet in spite of all these "nos", our empirical experience of the macrocosmic world proves that there is something here which completes the pattern of matter, the *sine qua non* of our existence, seeing that the electron holds the negative balance to the positive proton, making the atom as a whole neutral. Being neutral it is not one thing or another but can be everything equally. There is as a matter of fact no place where it is not—except in the empty spaces between.

The main point in all this is, for me, that the *continuous motion* of one of the fundamental factors in the universe—that which Oppenheimer calls "the heraldic symbol over the gateway of the new world", can be described, but never defined. I think his definition by negation brings us nearer to an understanding of what is in question than anything else. Perhaps electrons belong to that "reality" I am seeking, to which the measure "is" or "is not" does not apply?

VI

HAVING NOW LEARNT A LITTLE ABOUT the nature of the hydrogen atom, I think the next step should be to find out how proton and electron interact, for, although apparently electrons have nothing to do with the *structure* of the nucleus, they are evidently "locked together in an electrical embrace", as Lyttleton remarks, and together they are responsible for the whole physical world, from stars to man—that world of movement I am venturing into.

We learnt above that the inter-galactic gas is mostly composed of hydrogen; the Dutch astronomer, van der Hulst, discovered that this may emit radio signals even in the "waste-lands of space". Considering the hydrogen nucleus and its single electron (two billionths of an inch apart) revolving round each other at 600 miles per second, it seems little short of miraculous that a system so minute could project any kind of signals through millions of light years, becoming a radio transmitter. Just as do particles, proton and electron spin round mostly in the same direction. But under certain circumstances: "the electron may flip over into an 'upside-down' position so that it and the nucleus are spinning in opposite directions. When that happens a radio signal is generated. The flip over of the electron has the same effect as an oscillation, creating a ripple in space or a radio-wave".[1] It seems clear that without the inter-relationship of electron and proton, which is disturbed thereby, the "flip-over" of the electron would leave no mark. Furthermore, considering optical astronomy, Pfeiffer says: "Until recently the (interstellar) gas could only be seen in certain places near extremely hot stars such as those in the Trapezium star group in the Great Nebula of Orion. In these hot spots hydrogen is 'ionized'—it looses its negative electron, bears a positive charge and can be studied because it radiates light when it recaptures an electron", (showing the vital importance of this relationship!) Furthermore, as he points out: "If a sufficiently sensitive instrument scanned the spaces between the stars, it would pick up signals from hydrogen atoms and nothing else".[2] What causes these important "flip-overs" is at present un-known, they seem to be spontaneous events, though Pfeiffer says collisions may be responsible for particles taking on their former spin and starting the process over again.

Seeing the near emptiness of interstellar space where, according to this author "a hydrogen atom is an average of about 3/8 of an inch from its nearest neighbour", (in human terms about 48,000 miles) and that hydrogen atoms travel very slowly between stars, three miles a second is, as he says "snail's pace" in the atomic world, it is amazing that collisions can occur at all, in any case, they are extremely rare since they

[1] J. Pfeiffer, opus cit., p.156. [2] Opus cit., p.157.

only occur about once every 233 years according to the latest estimates. How then is it possible that the waves picked up by Radio telescopes appear to be continuous? Difficult though it is to visualise, there are apparently in the immensity of space such vast amounts of hydrogen atoms compared to our minute Earth, that they are continuously bombarding us from all directions and distances. I must admit that all this makes heavy demands on the imagination of those who are not astro-physicists, but I have limited myself to such details as seemed necessary to show something of the atomic activity all around us, and of course, equally in ourselves. Even though it would seem to make no appreciable impression on the human senses, we can be sure it happens all the same.

Speaking of hydrogen radiation, and considering the fact that electron waves and light waves have so much in common, it makes one wonder if on top of all its other functions, the electron is not the *source* of life itself? Early in this century, Einstein made a discovery about light that proved to be "revolutionary" for an understanding of atomic systems. That discovery led him to the conclusion that: "Light which we know as a continuous phenomenon propagating from point to point and from time to time, like a wave, is nevertherless also corpuscular, consisting of packets of energy determined by the frequency of the light and of Planck's constant"—the quantum h.[3] I should point out that "packets of energy" are still packets of motion, not of matter. I would rather say that their "corpuscular" aspect is an aspect of condensations of

[3] Oppenheimer, p.46.

motion. Furthermore if we consider the close relationship light waves bear to electrons, being, as they are, dependent on the relative spin of proton and electron, one is tempted to ask what relation this bears to creation itself?

Perhaps, we may find a tentative answer in relating some facts of physics to what we have learnt of astronomy, as R. A. Lyttleton has done. He has carried the whole question an important step forward, it seems to me, by his suggestion that we may be living in an "electrical universe", and by showing the *cosmic* effect of the interaction between the elements of hydrogen, and the decisive role therein of the electron. To make his ideas at all comprehensible however—they being an entirely new approach—I shall have to quote him rather extensively, but if we are to understand something of the true nature of the electron and the part it plays in the universe, I cannot omit it.

We have learnt that proton and electron are positive and negative partners in the atomic system, their charges having been supposed to be exactly equal. But according to Lyttleton: "if the proton has slightly greater charge than the electron numerically, then, instead of cancelling out to zero total charge, the proton and electron will give a hydrogen atom a slight positive charge—excess . . . If now the whole universe is thought of as a highly rarified cloud of hydrogen, then the charge-excess would produce a kind of electro-static pressure, urging every atom away from every other atom, and overcoming the attractive tendency that gravitation alone would otherwise give". Whatever theory we may prefer about the creation of the universe, we have to accept the fact that hydrogen atoms have existed from the beginning, whatever that may mean. What may then happen in the dense atomic condensations in the interiors of stars where that excess charge must be greatly intensified? We find that, as the excess charge overcomes gravitation, the material world would disperse instead of pulling itself together. In fact Lyttleton asks: "How then can galaxies possibly form? The only condition that could differentiate one part of space from another would be if the material anywhere became ionised—that is if the electrical bonds holding protons and electrons together as hydrogen atoms should become broken",—which does actually happen when sufficiently high temperatures have been reached. Stars for instance may heat the surrounding clouds of hydrogen to the temperature when ionisation sets in, separating the atoms into their constituent protons and electrons. "Once proton and electron become free particles, the charge-excess acts violently on them, tending to repel the proton out of any region and holding the free electron . . . so in any ionised region of space, there will always result just such an excess number of electrons as would offset the slightly greater charge of the proton. Every element of volume would thereby be rendered strictly neutral electrically. Once this happens, gravitation alone is operative, for the charge-excess disappears, and in such regions condensations into galaxies can occur and continue".[4]

What happens further on the ascending scale of creation? Pursuing our path of movement without

[4] *Rival Theories in Cosmology*, Pub. 1960.

substance—or exchanges of energy—Hermann Bondi says: "If the famous equation $E=m^2$, linking the mass of the body and its energy, means anything, it means that out of a certain amount of mass one can only get a limited amount of energy . . . A body of the size of the Earth is considerably compressed and squashed. Material in the centre of the earth has perhaps 50 per cent extra density over material at its surface of the same composition. In the Sun this effect is much greater still . . . If we go to still larger stars, then the force of gravitation becomes yet more powerful compared with all other forces, and if we imagine an enormous body of the mass of millions of stars, then gravitational forces in it may gain entirely the upper hand; and it may be in collapses of this kind, where gravitation has won completely over all other forces that we see the great energy sources which we suspect in the Quasars, those mysterious sources of radio-radiation that seem to have such a tremendous output of energy".[5]

Does this not suggest that in gravitation, (a "universal force which extends everywhere") lies some power that is not a "thing" anymore than is movement, co-ordinating movements in the universe up to the point where creation may start again? And thus, with gravitation we come a few steps nearer to what I am trying to explain, for gravitation has not even such "mass" aspect as have atoms of hydrogen, it is purely a force, and if I have succeeded in making it clear that atoms are of the nature of insubstantial movement, how much more so is gravitation? Is it not something in the nature of Mind, that Mind which the Lankavatara

[5] See *Listener* 27/1/66, also Fred Hoyle: *Galaxies, Nuclei and Quasars.*

Sutra claims is all there is. We are in fact re-discovering what the Chinese Third Patriarch, Seng-ts'an knew 1,300 years ago: "Being is an aspect of Non-being, Non-being is an aspect of Being." But it is clear that we would never know anything about Non-being (or emptiness) were it not crossed with Being, or that "order of events" of which Einstein spoke. Our *awareness* of either is subject to the existence of the other. "For distance like time, (emptiness like movement) is a relative concept and there is no such thing as a space interval independent of the motions of the systems of reference".[6] As the Tibetans realised long ago: "Everything that exists, exists in relation to something else".

I think we may now accept the fact that stars and atoms are systems in motion, in the final instance: motion itself, but at the same time we have to recognise today that the motion of the electron is still not the smallest of cosmic "events"; the so-called "fundamental particles" which seem to form the body of the proton (if we may call it so?) being much further down the scale. And even that may not be the end. There is a suspicion that those particles may even be connected in some way that is still a mystery with the beginning of all things. Speaking of the expansion of the universe which Lyttleton calls "one of the most sublime and perplexing problems in the whole realm of modern science", he says: "It is now coming to be recognised that it is in some way connected with the creation of the universe, and this in turn must be in some way connected with the fundamental particles that go to

[6] Lincoln Barnett, opus cit., p.46.

make up matter itself"[7]—as for instance enzymes go to make up human bodies! Little is known about these particles so far, however, and intensive study is going on in all countries. The only thing we can be sure of at present is the fact that nothing at all would be known of them had not their "spin" or motion been detected. In fact "particle" is probably the wrong word, for it conveys a material quantity to our minds, whereas we are not actually aware of anything but a disturbance (perturbation)! Following which, through electro-magnetism or gravitation, condensations take place. Collisions, decays, magnetic fields, cosmical repulsions, all influence these motions, draw them on or scatter them, it is all one: the flight in one direction is diverted into another, the speed is the only thing that counts, and that speed is something hardly conceivable, it would seem impossible that such a thing could be apprehended "The scattering cross-section or probability for positive pions begins to increase sharply at 100 million electron volts, (MEV), reaches its peak of somewhat more than 200 MEV and then falls almost sharply . . . Ordinary spin and isotropic spin then are two of the most important properties of any particle or resonance particle".[8]

It is almost impossible to imagine such things, but I think Pfeiffer's description opens our minds to a new world: "Matter in motion is universal and so is radiation . . . We know of no object which exists naturally at absolute zero (not even ice! LCB). Or, to put the same idea in different words: whose atoms

[7] *Illustrated London News*, 2/5/59.
[8] Prof. Hill: Resonance Particles, *Scientific American*, January, 1963.

do not move. Every object is at some temperature, has some atomic activity and emits some radiation".[9]

Is that borne out by our common experience? In gases, liquids or solids for instance? Apparently it is. In a gas, the molecules are in a state of rapid and irregular motion, but equally there is emptiness to be found: "They are separated by relatively large distances —on an average anything from 100 to a 1,000 times their diameter, but, because of their random motion they are continuously colliding with each other . . . The average velocity of the molecules is some 1,300 ft. per second (900 miles an hour), and the number of collisions to which *each* molecule is subjected in one second is approximately 6,000 million".

Liquids are in a similar state of agitation, each drop of water contains several thousand million, million, million atoms, as Eddington remarked, and, as we know, every one of these is in movement, but the molecules are much closer together than are those of gas, so the main free path is not much more than the diameter of the molecule. "In the third state, (solids) the molecules are so closely packed as to be unable to move freely, and we say that the material is solid. The molecules are still in a state of vibration, but their internal fields are strong enough to prohibit independent movement".[10] Frequently however they "co-ordinate themselves into structures of considerable symmetry which we call 'crystals'",—one of the wonders of nature.

We thus discover that motion is inherent in all we

[9] Opus cit., p.139.
[10] J. H. Reyner: *The Universe of Relationships*, pp.43-44.

see as well as in what is invisible to the naked eye—even in things that seem completely solid. In Tibet they seem to have known these facts long before our scientific age, for they held that: "The Universe *is* movement, and this movement is made up of contacts. (In our terms: collisions LCB). The contacts and their effects *are* the universe".[11]

If our physicists are now bringing an ancient truth to light once more, we may suspect that it is something that belongs to a secret knowledge of "reality", embedded in the unconscious mind of man.

[11] A. David-Neel: *Les Enseignements Secrets dans les Sectes Bouddhistes Tibetaines.* Translated by the author.

VII

POSSIBLY IT MAY BE THOUGHT THAT I HAVE said enough about this matter, but I want to ask my readers to have patience with me a little longer as I tell the story of the neutrino, for I feel it to be the most momentous of all the manifestations I have attempted to define thus far.

To begin with we may accept it as a fact that the break-up of nearly all particles is accompanied by the emission of neutrinos, in fact the neutrino seems to be at the bottom of all processes of the universe. Even when the neutron breaks down into proton and electron, "the mass of the neutron exceeds the mass of the proton and electron by the energy equivalent of 780,000 volts, a very tiny amount as these matters go—but which cannot be ignored all the same. As a result the neutron ought to resist beta decay. It does. Its life-time is about 18 minutes, by far the longest of any unstable elementary particle on record".[1]

That excess energy should be carried away by the energy of motion in the kinetic energy of decay products. But all measurements of these proved beyond a doubt that the proton and electron always carried

[1] S. B. Treiman: The Weak Interactions, *Scientific American*, March/59.

away less kinetic energy, never more. This fact set physicists a very difficult problem: either they had to accept the breakdown of the conservation of energy, one of the fundamental principles of physics, or, "they had to suppose the existence of a new unseen particle . . . Such a particle would have to be a neutral particle that should fly off in beta decay if energy was to be preserved, and there was every reason to suppose it could have no mass". No wonder it was unseen; all the same, whatever it was that could remotely be thought of as a "particle" it was named "neutrino". Investigations into its properties have gone on ever since it was discovered by W. Pauli and Enrico Fermi in 1931, but it is still one of the greatest mysteries of the physical world. According to the latest ideas on the subject: "Neutrinos are the most impalpable of particles. They have no electric charge, no mass (or none that has yet been measured) and they travel with the speed of light. They are produced in huge numbers by nuclear processes inside the sun and other stars. Those that encounter the Earth pass through it with ease. Only about one neutrino in every ten billion (10^{10}) passing through the centre of the earth is likely to react with another particle".[2] As all elementary particles reveal their presence by interaction, "obviously a particle that reacted with nothing whatever, could never be detected". (*ibid*) All that can actually be "known" of this "wondrous" particle is that it moves in a straight line with the speed of light through any container we choose to think of. The neutrino would seem to

[2] Leon M. Lederman: The Two Neutrino Experiment, *Scientific American*, March/63.

confirm Galileo's First Law of Motion? Is it perhaps motion in essence?

Experiments with it were made with the great Brookhaven synchrotron which has a shielding wall of 13.5 metres of steel armour plate from an old battleship. "The steel wall stops all the particles except the neutrinos which penetrate the wall as if it were not there. The pions and other strongly interacting particles penetrate only about a foot before being stopped . . . When the synchrotron was well-behaved, ten million neutrinos passed through the spark chamber of each pulse. In a good hour the machine delivered 3,000 pulses and a good day had 20 good hours. The experiments ran for 25 days during which time nearly 10^{14} or 100 trillion high energy neutrinos travelled through the spark chamber. That is about the number of low energy neutrinos from the sun that pass through our bodies every second producing perhaps one reaction in a human life-time."[3]

I make no apology for this long exerpt; the question of the existence of the neutrino and its activities appears to me to be of prime importance for some unknown reason. So important is it that scientists have not rested content with the above experiment, but have continued to make attempts at detecting the nature of the neutrino all the world over. A group of Indian, Japanese and British investigators are at work in a gold field near Bangalore, 7,600 ft. below ground. At a mine in South Dakota, Raymond A. Davies Jn. of Brookhaven is engaged in building a giant radio-chemical detector in hopes of detecting the flux of

[3] Ledermann, opus cit.

electron-type neutrinos from the sun. Then in 1962 a group of physicists from the Case Institute of Technology in the U.S.A. decided that an underground arrangement would be simplest provided that a deep enough site of adequate extent could be found. A world-wide search resulted in the selection of a location in the East Rand Proprietary Mines near Johannesburg. There physicists have dug down through 10,492 ft. of solid rock and built themselves a laboratory below ground. The Case group has been joined by investigators from the Witwatersrand University and the work of the laboratory is supported by the U.S. Atomic Energy Commission, which would prove, if nothing else did, how important the identification of the neutrino is considered to be. One cannot but ask oneself why all these millions should be spent in the detection of a particle that would not appear to be of any practical value? Governments are usually not very fond of spending money without any appreciable return! Men seem to be haunted with the idea that there is in this strange phenomenon a new aspect of "reality"—perhaps of Mind itself?

What is actually known about it thus far? How is it produced, what does it do? "Low energy neutrinos,—in the range from zero electron volts to roughly 15 million electron volts—arise predominantly from the steady nuclear fires believed to power the overwhelming majority of the stars. Some neutrinos in the low energy range or perhaps a little higher, may be produced by nuclear reactions that occur in the explosions of super-novae. High energy neutrinos—ranging up to tens of thousands of billions electron volts—originate from at least two sources. The mechanism by which neutrinos are produced in the earth's atmosphere by collisions of primary cosmic rays (mostly protons) with atmospheric atoms is fairly well-known and probably predominates . . . Less is known about the second source of high energy neutrinos, but in a way they are even more interesting. These neutrinos are produced by collisions between cosmic ray protons and matter in the vicinity of stars and in intergalactic space. In addition it is conceivable that neutrinos with various energies may have been produced in the reactions that accompanied the beginnings of the universe and that these neutrinos may continue to wander through space and time".[4]

I do not need to go into the question of the electron-neutrino and the muon neutrino because what interests me is not its mass, which is probably zero in both cases, but its motion. It is a sobering thought that of the 10 billion neutrinos that pass through every square meter of the earth's surface every second, only about ONE is absorbed by the earth! Philip Morrison says that: "perhaps 6-8 per cent of the total energy released by the sun is in this transcendental form, passing without change through star, space and planet; even at night neutrinos stream through us from the hidden sun, travelling right through the massy earth as though it were not there. The neutrino flux from the sun carries 40,000 times more energy than moonlight, yet we have never heard, seen or felt its presence".[5]

[4] Fred Reives and J. P. F. Sellschop: Neutrinos from the Atmosphere and beyond. *Scientific American*, February, 1966.
[5] The Neutrino, *Scientific American*, January, 1956.

This is a truly fantastic tale, hardly credible, yet I had to tell it for it is the most complete expression of what I mean when I speak of "motion only". The neutrino *is* motion without substance. If, in giving it a name and talking about "it" we imagine it to be some thing, we are far from the mark. Prof. Frisch says: "The neutrino seems to be nothing but a *disembodied spin.* It has no electric charge, no magnetic effect, no mass", which would explain why, although its tracks can be discovered, it can never be stopped. Anything which can be stopped must have been a moving *thing*, but the neutrino never was a *thing*, only a perturbation in space, motion itself! When Morrison calls it a "transcendental form", and Frisch a "disembodied spin", in either case the neutrino is not what we have been accustomed to regard as "matter", in fact it is about as far from any material object as the human mind can conceive of, the bottom so to speak. It rushes through space unbeknownst to us till it makes one of its rare encounters, but the fact that it always appears with beta decay begs the question: where it comes out, must it not have gone in? Wherever the parity principle common to the whole atomic world, breaks down, i.e. in the world of the fundamental particles, the ultimate state of material phenomena so far known, "there is always at least one neutrino amongst the decay products". If in its final state, when matter is reduced to a mere "spin", something transcendent, namely the neutrino, is emitted in the breakdown of the nuclear process—equally the capture of the neutrino might be the beginning of everything? Prof. Peierls says: "If it is possible for a neutrino to be emitted in a nuclear process, the converse must also be possible, namely that a neutrino colliding with a nucleus would be absorbed, either together with one of the atomic electrons which are in the neighbourhood or leading to the production of a new electron".[6] And from there up the scale to the greatest nebulae and galaxies. New atoms would continuously be created throughout the universe, but their beginnings could never be observed.

This idea does not run counter to the pattern of nature with which we are familiar, always the largest grows out of the smallest: stars out of interstellar dust, Sequoia out of seed, man out of chromosomes—only to fall back eventually by *decay processes* (in other terms "beta decay") into the smallest once more, to become manure for the new birth. Being all weak like the seed, and only manifesting in weak reactions, the probability is that the neutrino will never do anything but act as if it were in a vacuum. It rushes past nucleii in the time order of 10^{-23} seconds, and not only that but past very many without paying the slightest attention to them, there is in fact only one in a million chances that it may meet with any obstacle on its path,—yet we have to consider that one part. Not every seed that it sheds grows into a Sequoia—the greatest tree on earth—yet one amongst its millions of seeds does. Treiman says: "Once a neutrino is produced, it heads off to the ends of the universe and, *as far as we know*, does no more".[7] But is that really the end of it? Speaking of the "creation out of nothing", Bondi and Gold have calculated that the creation of matter must

[6] Opus cit., p.247. [7] Opus cit. (italics mine).

proceed at the rate of one hydrogen atom per hour per cubic mile in galactic space. Could this rarity in the creation of hydrogen atoms perhaps be connected with the rarity of stopping the neutrino? (See note).*

Prof. Peierls seems not to exclude the possibility, for he admits that: "In spite of tremendous odds, an indication of the inverse beta process has now been found".[8] I should much like to know what that "indication" is, for I must confess that all I have been saying looks very much like guess-work—but guess-work has sometimes been at the root of new discoveries. A universal spin-movement, no one can now deny; it may be that *movement in nothing* is the source of all life? We see it now as the Omega, should it not therefore also be the Alpha?

And so we re-discover what wise men in Tibet knew several millennia ago, probably we and they, all equally draw our knowledge from out of that collective unconscious Mind, container of all wisdom. There is no reason why we should not become aware of the same manifestations as our ancestors recognised, for such knowledge lies forever within the nature of man, ready to be unearthed. If we ignore some aspect of it, (as our ancestors ignored electricity) that is only due to our blindness. It only lies with us to discover that: "The tangible world *is* movement not a collection of moving objects, but movement itself".[9] Just a disturbance in emptiness! OR, a stimulus to Fullness.

* To my mind the new idea about the "bubble" does not alter this question.

[8] Opus cit., p.247. [9] A. David Neel, opus cit., p.30.

VIII

OF COURSE THERE ARE MANY OTHER manifestations of movement in the universe, and before I turn to the question of emptiness, I should like to consider what forms it takes in our own bodies. Eddington says: "Nearly mid-way in scale between the atom and the star is another structure no less marvellous: the human body. Man is slightly nearer to the atom than to the star. About 10^{27} atoms build his body; about 10^{28} human bodies constitute enough material to build a star . . . As regards mass, man is rather too near the atom and a stronger element for the midway position would be the hippopotamus".[1] But if Eddington had known what we know today about the fundamental particles, he might have advanced us to the midway position!

To begin with: what about these human bodies: are they at rest even when apparently so? We know of course that they are made up of molecules in motion, but let us just consider one aspect of this which is typical of the interplay of the whole. Enzymes, for instance. "Enzymes are molecules designed to act as catalysts, i.e. to assist in chemical transformations

[1] *Stars and Atoms*, pp.9 & 85.

without becoming involved themselves. This they do by entering into relationship with another molecule, such as an amino-acid and forming a bond with certain of the atoms. This releases the tension in some other part of the structure which then breaks up into a different combination, after which the enzyme lets go and goes on to call on some other molecule. A relatively small number of enzymes is sufficient to promote various transformations. But the enzymes themselves are also found to be dependent for their action on still smaller molecules called co-enzymes, the important constituents of which are the chemical substances called vitamins".[2] Further, considering the amino-acids which are the constituents of all living cells, "with only incidental exceptions, the same 20 amino-acids are found among the proteins of all terrestrial organisms; furthermore all have a left-hand twist although in principle a right-hand twist would have been equally suitable".[3] Here we seem to obtain the same kind of rule that we have found throughout, and what is of vital importance to all life on earth is, that: "nucleic acid provides the means by which life propagates itself from particle to particle from generation to generation in all living things from the smallest particle of virus to the largest Banyan tree".[4] All equally are part of the same law of motion, regardless of the kind of physical "bodies" resulting therefrom. And that motion is what conveys the idea of "life" to our minds, though the difference between

[2] Reyner, opus cit., p.19.
[3] Prof. Joshua Lederberg. See *New Scientist*, 18/2/60.
[4] C. D. Darlington, *Listener*, 21/7/49.

what is "alive" and what is not, is shrinking more and more.

But even amongst living things the motions which give rise to the differences in their nature must have very different velocities: the difference in the rate of circulation in the sap of the Banyan tree and that of our blood-stream must certainly be reckoned in hours if not days, and even faster than our blood-stream move the enzymes.

But that is not all: there is the biologist's micron, usually called "mu". A mu is a thousandth of a millimeter or about a 25,000th of an inch, of course only visible through a microscope, magnifying it about a thousand times. Michael Abercrombie tells us that the mu world "proves to be as full of interesting and beautiful shapes and patterns as any part of nature. It is mostly liquid jelly, so that the shapes are mostly curves . . . There is plenty of movement in and amongst the cells, usually of the gently flowing kind usually associated with ghosts, though it is also possible to find vigorous exact rhythmical movement".[5] After all that has been said, this is only what might have been expected, but Abercrombie gives a final picture that is most exciting and unique in my experience of such things. None of the motions we have investigated thus far was entirely inconceivable but this is something fantastic. He says: "It won't be long before we are aware that this world (of the mu LCB) has a phenomenon which our normal world has not; small particles any less than a few mu in diameter, when they are floating in liquid are never still. They jerk and

[5] See *Listener*, 18/11/59.

shiver and twist, and the smallest positively dance about". (*ibid*). That is going on inside our own bodies all the time and the only thing I can compare it to is the Hindu idea of the creative dance of Shiva—the mu's dance is possibly also creative? That would be another example of the intuitive understanding of ancient sages.

IX

I HOPE THAT IN ALL THIS I HAVE BEEN ABLE TO convey something of my awareness of living in a world, not of matter nor of things, but of impalpable waves travelling everywhere, every wave affecting every other, universally and infinitely. As John Pfeiffer most graphically describes it: "Imagine a spider-web stretching from here to the edges of space and time, from here to eternity. The web is a vast detection net-work, a kind of inter-galactic burglar alarm system. Let anything happen anywhere, let some imaginary fly, about as large as the Milky Way, get its wings caught in the web and the whole cosmos knows it. The web trembles all over . . . Ruffle the universe anywhere and disturbances of energy spread throughout the space between the stars. Many events affect the cosmic webs. Here two spiral galaxies crash head-on. Here a star bursts and here are great jet streams, and here whirlpools or vortexes of gases. Nothing can happen in the universe without making its mark, without signalling the fact of its own happening—and the signals are radiation".[1] Who then can doubt that all is ONE?

[1] Opus cit., pp.137–8.

In terms of energy of living forms, Sherrington expresses the same fact: "If we think of it (the living energy system LCB) as an eddy in the stream of energy which tends to grow; as part of its growth we have to reckon with its starting other eddies resembling its own",[2] and so on *ad infinitum*. If we accept these facts and accept ourselves as some kind of "fly" in that vast web, then everything that happens in the universe must also leave its mark on man. And for such marks, the first place where we may look to find them is undoubtedly in our brains, where there are greater signs of evolutionary advance than in any other living phenomenon. As Eddington remarked: "The story of evolution ended with a stirring in the brain organ of the latest of Nature's experiments; but that stirring of consciousness transmutes the whole story and gives meaning to its symbolism. Symbolically it is the end, but looking behind symbolism, it is the beginning".[3] We have often regarded brain-waves metaphorically, but now we shall have to admit reality behind the metaphor. They are as "real" as any of the other waves we have been considering; the brain is in fact the most wonderful and complicated system in nature, and I may add, the most difficult to decipher.

We have learnt a great deal about the activities inside the brain in this last half century, possibly no science has made such vast strides as this, but even so, all the wisest physiologists and biologists agree that the biggest problem, namely how the brain controls our lives, is still unsolved. Therefore I can only give such

[2] Sir Charles Sherrington, *Man and His Nature*, p.78.
[3] *Science and the Unseen World*, p.25.

facts as are known at present, not in any way presuming to assert that this is the last word. We can measure the frequencies of the nervous system, but to interpret how these produce the image we construct of the world around us would be quite a different matter. As Sherrington said: "it is not nerve impulses which are likely to tell us what we want to know about the brain. They are a concern of the nerve fibres. Nerve fibres are merely the wires to and from the telephone exchange. We want the activities of the exchange itself".[4] The "exchange" is a system of nerve cells and we are told that inside our heads there are about 10,000,000,000 of these. "This is not a very large number as living cells go . . . but unlike the cells in your blood or skin or muscles, the brain cells are in communication with one another. As a result the number of ways in which they can interact is vastly more than the number of electrons in the whole universe!"[5]

How this communication is effected is of course a very difficult thing to establish, but great strides have been made thanks to the electro-encephalograph by means of which rhythmic waves have been discovered in the brain. Sherrington says that: "the brain-waves are oscillations of potential, recurring 10 times a second . . . The waking roof brain is never free from electrical action".[6] The roof brain, in other words the cerebrum, which we commonly think of as our heads, should be distinguished from the cerebellum, the

[4] Opus cit., p.219.
[5] W. Gray Walker, Mirror of the Mind, *Listener*, 1/10/59.
[6] Opus cit., pp.202-4.

"lesser brain", which lies at the back of our skulls upon the brain stem, under the over-arching mantle of the great hemisphere of the cerebrum. According to Sherrington: "the roof-brain is the nervous organ which *par excellence* can learn by experience and it deals with final causes." (It is probably what I am using at this moment). And the brain-waves may be regarded as "the *patterns* of action set up among the millions of cells in the brain",[7] on which J. Z. Young says we must concentrate our attention!

But we must realise that cells themselves are no more a "thing" than any other material in the universe. "The cell is a unit of life. A speck of material which is said to "live" while the vast majority of specks of material are regarded as lifeless. Has it some particular element of matter in it which those other specks have not"? Sherrington asks. "No, that is not the key . . . The elements of matter . . . in the living cell are among the very commonest of those spread broadcast in material which does not 'live', in soil, rock, air, and water . . . But in the speck that lives, the common elements are differently compounded . . . they compose a tiny droplet of granular jelly. That is, *it is not material in any one single pure phase*—solid, fluid, or gaseous; it is a mixture of all these". (Italics mine). Although fundamentally immaterial, that jelly was recognised as "the physical basis of life", to substantiate it, we call it protoplasm.

But a change has been brought about in these ideas by the recent advances in biology. Until now chemistry had been dealing with "crystalline stuffs" but, as Sher-

[7] *Doubt and Certainty in Science*, p.60.

rington points out: "Perhaps they, (the chemists) had not so often, as had the rank-and-file biologist, lingered over the microscope and followed the mysterious *dance* (again!) of the nuclear segments about the astered spindle in the living cell. Though still microscopic it is still vastly larger than any molecule . . . There would be room on the surface of one cell for some hundreds of millions of the great rod-shaped or skein-shaped protein molecules bristling with chemical receivers. Its content of these rushing, whirling, electrical charges, some of them spinning many millions of times a second, is beyond practical enumeration. Having no parts, being structureless, and therefore unable to come to pieces, it was immortal".[8]

Here once again we discover movement, nothing else. A spinning and dancing in the body as everywhere else—co-ordinated by the electrical waves from the brain.

One last word about the cerebellum which is perhaps the most puzzling thing in the whole human make-up. Though it is obviously of vital significance, about what it *does*, there is so far no agreement. "The cerebellum generates electrical waves at the highest frequency in the nervous system. They range from 200-400 per second—ten times faster than those of the cerebral cortex. (But on the other hand) "the amplitude of cerebellar discharges seldom exceeds 200 micro volts; this is a fraction of the operating level of the cerebrum, which runs 10 or more times higher". Yet it has been found that: "the cerebellum imposes its peculiar high-frequency, low amplitude characteris-

[8] Sherrington, opus cit., p.69.

tics upon all the impulses that pass through it", that is, all the nervous impulses of the body. But what purpose is served by this peculiar activity? "Some suggest it may be the 'great modulator' of nervous function; . . . or it may act as a 'band-pass filter' that is, its fast activity may allow some wave-patterns to go through whilst blocking or altering others".[9] Another suggestion is that the important thing is that the total energy added by the cerebellum to the outgoing impulses makes of it the arbiter of our fate, for it designs the patterns our brains construct. That would confirm the idea held by many that its major function is the integration of all activities of the body! That would make its function as "transcendental" as those of the neutrino—and as vital to the whole.

Seeing that all the facts I have given about our human condition are taken from what we are accustomed to regard as "life", I think we must agree with Sherrington that: "The term 'life' has no boundary from 'lifeless', since all equally is movement". But we are still left with the question: where lies the connection between brain and mind, that MIND which the Lankavatara Sutra says is all there is?

I should like to give the replies of three men, all prominent in their approach to these questions: W. E. Le Gros Clark, an anatomist, Dr. Penfield a neurologist and Sherrington. The first says: "No more than the psychologist is the anatomist able even to suggest how the physico-chemical phenomena associated with the passage of nervous impulses from one part of the brain to another can be translated into a

[9] Ray S. Snider: The Cerebellum, *Scientific American*, August, 1958.

mental experience". The second says: "Thus I am suggesting that the master motor area . . . may be found at the level of the upper brain stem"; but he ends by asking the basic question: "What is the real relationship of this mechanism to the mind?" And clearly Sherrington himself seems to have been dissatisfied with the answers, for at the end of it all he admits that: "we have to regard the relation of mind to brain as still not merely unsolved, but still devoid of a basis from its very beginning".[10]

It seems to me that Erwin Schroedinger comes nearer to the idea of the Lankavatara than anyone else I have met with. He wrote: "Mind is by its very nature a *singulare tantum*. I should say the over-all number is just one . . . namely mind is always *now*. There is really no before and after for mind".[11] Where is the difference between that and: "As far as the duality of Being and Non-being extends there is the realm of intellection; when this realm vanishes, intellection completely ceases . . . Of neither existence nor non-existence do I speak, but of Mind-only".[12]

[10] The Physical Basis of the Mind. Talks, pub. 1950.
[11] *Mind and Matter*, p.62.
[12] Lankavatara Sutra Chapters LXII & LXIV.

X

WE SHALL NOW HAVE TO CONSIDER THE sources from which the brain draws its information and constructs our lives, as far as this is possible in this short study. Primarily the information derives from the five senses, ("skandhas" in Indian terms) and here we are faced with movement of a different kind, namely vibrations. There can be no doubt that the brain emits and receives vibrations continuously to and from the senses, but I have reached the conclusion that our aural ones have peculiar qualities not found in the other senses. All religions from the most primitive to the most mystical have been influenced by *sound.* As Reyner truly remarks: "It can produce the mindless ecstasy of savage rhythm, (for instance with tom-toms LCB) or a state of exaltation . . . beyond the customary emotional level".[1] In the midst of the Amazonian rain-forests some Maquiritare Indians, who had never seen a white man before, could sit for hours entranced, listening to Mozart being played on Pierre Gaisseau's phonograph. And Bushmen told Laurens van der Post that they could hear the music of the stars, which

[1] Opus cit., p.93.

gift they seem to share with Pythagoras! Just vibrations but what effects! Is it possible that in this we may see a link between ourselves and the universal movement side of the Great Atman?

All concepts derived from other senses are, consciously or unconsciously, the results of attention. Sound surrounds us and reaches us from everywhere apparently, it is even cosmic—perhaps it affects our lives in ways of which we were unaware until now? (See Radio transmission!)

I think that the *influence* of sound is best illustrated by the "mantras" (words of power) used by the Tibetans since time immemorial. Lama Anagarika Govinda claims that: "The secret of this hidden power of sound or vibration . . . as it reveals the nature of things and of the phenomena of life, was well understood by the sects of olden times; the Rishis who inhabited the slopes of the Himalayas, the Magi of Iran, the adepts of Mesopotamia, the priests of Egypt and the mystics of Greece . . . Pythagoras, (evidently an initiate in the mysteries LCB) spoke of the 'harmony of the spheres', according to which, each celestial body—in fact each and every atom—produced a particular sound on account of its movement, its rhythm or vibrations".[2] If we accept this idea and listen carefully in places where we can be sure that there is no other sound whatsoever, we may hear the sound of these vibrations in our own ears. The Rishis of India are credited, to this day, with being able to receive waves of word-sound in their solitary caves, and there are well-authenticated cases of Indian fakirs

[2] *Foundations of Tibetan Mysticism*, p.26.

receiving messages from great distances. My father was present at one such an occasion in India when a message was received from London, which was later confirmed.

Lama Govinda quotes the experience of Madame A. David-Neel who on her Tibetan journey met a master of sound, "who not only was able to produce all kinds of strange sounds on his instrument, a kind of cymbal, but who—like Pythagoras—explained that all beings and all things produced sound according to their nature, and the particular state in which they find themselves. 'This is, he said, because these beings and things are aggregates of atoms that dance (!) and by their movements produce sounds'. When the rhythm of the dance changes, the sound it produces also changes". (We may remember what happens when the spinning atom turns upside down LCB). "Each atom perpetually sings its song and the sound each moment creates dense or subtle forms. Just as there exist creative sounds, there also exist destructive sounds. (The Word of God—the walls of Jericho!) He who is able to produce both can at will create or destroy".[3]

Of Mantras, Lama Govinda says that their effect depends on the spiritual attitude and may not even be a vocal sound, it is on another scale. "Just as a chemical formula gives power only to those acquainted with the symbols of which it consists and with the laws of its application—in the same way a Mantra gives power only to those who are conscious of its inner meaning, are acquainted with its methods of operation and who

[3] ibid, pp.26-7.

F

know that it is a means of calling up dormant forces within us, through which we are capable of directing our destiny and of influencing our surroundings". (*ibid*). I find in this the very essence of all I have been saying: we discover waves of sound or movement, capable not only of influencing our personal lives, and of being used by us to exert influence on the world around us, but which are also an expression of the particular nature of any time or people. I do not regard it as mere chance that Indians have quarter tones in their music which we have not, or that, in our times, Schoenberg evolved the 12 tone scale. These sounds not only exert influence but are influenced by the world of which they are an expression; when one listens to them one realises how perfectly they harmonise with the conditions out of which they sprang.

There are countless instances of the ways in which the waves of sound, colour, smell, etc. that reach their senses, influence the lives of all creatures. I feel sure that smell for instance influences human as well as animal reactions far more than we are aware of—every human being has a distinctive smell which probably influences the reactions of others to us—bees recognise that. But there are also waves higher or lower than those to which our ears are attuned; dogs for instance hear sounds to which we are deaf; birds hear worms moving in the ground—perhaps vibrations from conditions (temperature and others) in particular places, help to guide their migrations? Insects catch waves on their antennae; snakes which have no ears, react to music played by the snake-charmer on his pipe. Termites "see" without eyes. But I have said enough about effects of sound waves to enable my reader to discover what he may of the rest for himself.

But I must insist on the fact that all vibrations of whatever description received by our senses, could never be assimilated by us, were the senses themselves not made up of cells in motion, directed by the currents from the brain. "Photons or particles of light may be messengers between electrical forces,—mesons between nuclear forces"—vibrations between cellular forces.

One last word about this power of sound. I know that it probably concerns something one should not attempt to talk about, as a wise man pointed out to me when speaking of what is known in Buddhism as the Trikaya. But there is an aspect of one of the three members of it, namely of the Shambhogakaya, which I feel I cannot ignore in this context, whilst at the same time warning my reader that this is only *one* of its aspects. Evans-Wentz gives us an indication of what I mean which is highly significant: "Then dawneth the *Shambhogakaya* (the invisible, super-physical body-aggregate of the perfect spiritual attributes of a Buddha) experienced as sound", his usually accepted attributes being purity and bliss. This author comments on these words: "As form, the Shambhogakaya is mentally inconceivable, as sound it is not wholly beyond mental comprehension."[4] Let us therefore apply our "comprehension" thereto; some day it may lead us to undreamt of results.

[4] *Tibetan Yoga and Secret Doctrines*, p.213.

XI

FINALLY, BEFORE LEAVING THE MOVEMENT side of this study, I would like to consider something which is still a dark science, but as experiments in it would seem to depend on vibrations, I feel I cannot ignore it. I mean ESP. We assume for instance that it is impossible for me, here, to know of an accident which happens to a friend in Australia at the very moment when it occurs. Yet there have undoubtedly been occasions when this has happened and we have had to accept the fact without explanation. There are other cases when people have known through dreams or visions, things that have lain in the future. But if we can imagine that under certain circumstances there is some part of us capable of receiving waves travelling at almost the speed of light (as we see pictures on our television) we might conceive it possible that we could "see" things or receive some kind of waves from a distance, as it were, pushed forward onto a beach by the swell behind. C. G. Jung has written a book on Synchronicity, which I feel is something which covers things of the same nature, occurring at the same moment, caught up in the same wave of energy, and hence producing the same rate of vibration, regardless of the medium.

We should remember that time may be one of our illusions, bound up with the City of the Gandharvas; if we try to imagine the true nature of time measurements, they certainly fade into the Principle of Uncertainty! How can we hope for instance to grasp what it really signifies when we are told that the light from Cygnus A takes 270 million light years to reach us? If we try to imagine light travelling 186,000 miles per second,—six million miles a year,—270 million of these is inconceivable! And what about neutrinos, with energies higher than 10 billion electron volts? Our minds attach such figures to things in order to distinguish them, apart from that they make no sense. Eddington perfectly sums it up: "The vastness of the time-scale of stellar evolution is *less* remote from the scale of human experience than is the minuteness of the time-scale of the processes studied in the atom".[1] Obliged to consider periods of time which as he says "appal our imagination", we ask ourselves: perhaps there is neither time nor space? and so we face the challenge of J. W. Dunne's explanations. He went very deeply into this question of time and its unreality.

If it be as Dunne says that: "if time passes or grows, or accumulates, or expends itself, or does anything whatsoever except stand rigid and changeless before a Time-fixed observer, there must be another Time which times that activity of, or along the first time, and another Time which times that second Time and so on in an apparent series of infinity".[2] Thus you and I with our time measurement may "know" that it was last month when my friend was killed in Australia,

[1] *Stars and Atoms*, p.85. [2] *Experiment in Time*, p.124.

but the observer at Time 2, may be able to take in both Australian time and mine in *one*, and so on—Time behind Time, to infinity. "The days that were six or seven days ago, and the days that were 6,000 years ago, they are as close to this present day as is yesterday" says Eckhart.[3] Not only 6,000 but for an observer at the heart of the Milky Way, *all* Earth time is one, and is it not a fact that we may be seeing the light of stars that have long since disappeared? That makes time as much of an illusion as mass; they become aspects of one and the same movement.

I would call Dunne's Time that which cuts across our time norm—"perpendicular Time", in which we do not see a concatenation but a periodicity; we do not measure it from here to there, from last year to this, we see it as one—as I see that tree trunk as one, my glance does not have to measure it by inches. Into that time, our consciousness irrupts in patches, like searchlights onto a landscape, and Dunne puts the pertinent question: "How are we to know which is the present configuration? And what is the use of my telling you that you must put your chosen instrument at the *now* on the time map (I in my chair, my friend in Australia) before you have discovered where that 'now' is? The instrument, (my clock) may *mark* it when found; but since you can change the instrument and the object about at will, neither of these can *make* it."[4] What then can fix the moment *now*? If that cannot be done, then we only obtain momentary glimpses of the time landscape which we call the now. But Dunne asks a question which was asked long before our era by

[3] Quint Transcription, p.203. [4] *The Serial Universe*, p.209.

followers of the Buddha: who is it that is observing the now? To be conscious of the time being observed by A, there must be observer B, and to realise what B is observing of A there must be an observer C who takes note of what B observed of A, and so on to infinity so where is my "real" now? We are like travellers through lines of movement, our conscious awareness of anything is just a crossing of those lines; our "electron" is caught up, or "ionised" by a wave, gets free, and going on, gets caught up again, as long as life endures.

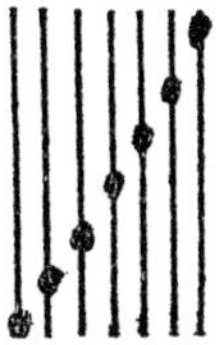

Dots being the time-traveller across waves of movement.

We imagine there is a before and after, but in reality there is none.

A friend had an interesting experience which I think well illustrates these facts in terms of something that may happen to any of us: she was sitting on a grass bank by the road-side in California one day, handling an old Chinese coin which she carried in her pocket as mascot. She suddenly felt something hard under her hand in the grass, and picking it up, found it was a coin the counterpart of hers! Together they had reached a "knot". Have we not all heard of thunderstorms, earthquakes, and the like, occuring at critical moments in history, or at the death of prominent men? I once asked Dr. Suzuki how he explains such co-incidences.

He said he did not think they are fortuitous but crossing threads on the road of life, not individual life, Life as such. In atomic terms: collisions.

In all this we get an inkling of an entirely different kind of movement, of which Eddington had a wonderful premonition: "From the window of our railway carriage we see a cow glide past at 50 miles an hour and remark that the creature is enjoying a rest"—is the cow resting or are we?

Thus anyone meeting such knots in time may at one moment know anything within its periphery, however distant it may be on a map. At present there is no scientific instrument capable of measuring this kind of time, but I do not think there can be any doubt but that it belongs to the universal law of motion. As D. H. Wilkinson pointed out: "We must admit that there are three great classes of action at a distance: gravitational, electric and nuclear which are the motive forces of the physical world, and which cannot in any sense be explained or even interrelated".[5] Should we perhaps regard the phenomenon of extra-sensory perception as a manifestation of one of these "actions at a distance"? Perhaps of the gravitational class, conditioned by waves set up by the senses?

[5] *Listener*, 3/5/58.

I

EMPTINESS

"Emptiness is beyond eternalism and nihilism".

LANKAVATARA SUTRA

I THINK I HAVE NOW SAID SO MUCH ABOUT the "movement" aspect of the "Great Atman", that, if my reader has not laid down this book long ago, having had enough of being told there is no substance in his familiar image of the world, he will have accepted the fact that there is no massive world at all. We see the wind in the trees, across the desert, and on ocean waves, but where there is none of these, we can still feel it blowing in our faces, although we *see* nothing. That applies to everything without exception, it is only the dimensions that differ. From the wind across space to the neutrino, all is but a perturbation in emptiness—or Mind? *What is, is but a movement in what is not.*

"The Form is the Void, the Void is the Form; Void is nothing but Form; Form is nothing else than Void".[1] Or, in scientific terms: "Distance like time, (emptiness like movement) is a relative concept and there is no such thing as a space interval independent of the state of motion of the system of reference".[2] Having studied one of these "systems of reference", we have

[1] *The Heart Sutra*, translated from the Tibetan by Lama Thongpa Tulku, and Getsul Lobsang Jivaka: *Middle Way*, May, 1965.

[2] Lincoln Barnett, opus cit., p.46.

now to discover what can be known of the space interval—of the "void".

Martin Johnson writes: "All views of atomic structure since Rutherford's first triumphs in 1910 agree on one point, namely that an atom is an extremely 'open work' structure—empty space over 999 thousandths of its volume and only held together by the electrical forces of that halo of electrons".[3] We may also remember what Peierls said about the emptiness of the solar system. Many years ago, Eddington expressed this fact very clearly, physiologically. He asserted that: "The atom is as porous as the solar system. If we eliminated all the unfilled space in man's body, and collected all the protons and electrons into one mass, the man would be reduced to one speck just visible with a magnifying glass",[4] the rest having been emptiness. But it is clear that we should never know anything of such empty spaces were there not what Einstein called "the order of events" filling it. How should we know anything of motion if all matter were a solid block as it must have been seen to be by the believers in the famous ether? Nothing could ever change, it would be a frozen world. But: "from it's (the Unified Field Theory's) august perspective, the entire universe is revealed as one elemental field in which each star, each atom, each wandering comet and slow whirling galaxy and flying electron is seen to be but a ripple or tumescence in the underlying space-time unity",[5] in what I call "no-thingness". Once

[3] *Astronomy of Stellar Energy and Decay*, p.70.
[4] *Nature of the Physical World*, pp.1-2.
[5] Lincoln Barnett, opus cit., p.98.

Einstein had rejected the ether theory, there was no tape measure by which to calculate the absolute motion of anything in the universe, and from that discovery came this fundamental postulate: "All the phenomena of nature, all the laws of nature are the same for all systems that move uniformly *relative to one another*".[6]—but not relative to space. I may know that one wave-length is longer than another, but not what its absolute length may be—perhaps that does not exist! Again to quote Eddington: "(nature) is not enthusiastic about frames of space . . . she herself has paid no attention to them". Motion has no frame, around it is emptiness, void, nothing, and this very emptiness is what makes all motion possible.

Of the fundamental structure of the universe, Einstein has pointed out that its highest edifices, (perhaps those "three great classes of action at a distance" of which Wilkinson spoke) "have been purchased at the price of emptiness of content".[7]

Thus the greatest of modern scientists is at one with Lao-tse:

> Thirty spokes unite in one nave,
> And because of the part where nothing exists we have the use of the carriage wheel;
>
> Clay is moulded into vessels,
> And because of the space where nothing exists we are able to use them as vessels.
>
> Doors and windows are cut out in the walls of a house,
> And because they are empty spaces, we are able to use them.
> Therefore on the one hand we have the benefit of existence,
> and on the other we make use of non-existence.
>
> Tao-Tê h-Ching Ch. XI.

[6] Ibid., p.28. (italics mine). [7] Lincoln Barnett, opus cit., p.100.

No scientist has ever stated the situation in the universe more clearly!

It is because of that emptiness that Lyttleton can say: "matter is still being created in the vast cathedrals of space". And even though the idea of "continuous creation out of nothing" is now being questioned, this does not make those words any the less true, for equally if we accept either of the other solutions of the problem: a stupendous explosion out of a super-condensation of matter, 5–8 billion years ago, from out of which the universe has been expanding ever since, (which doesn't answer the question what was there before or around that condensation); or Fried-man's solution of Einstein's equation which results in a pulsating universe, with alternating expansion and contraction; or even Hoyle's latest idea of a gigantic "bubble",[8] this does not change the fact that some-where at some stage or other there has to be emptiness. Long before Hoyle took up the idea, Sir James Jeans remarked: "The substance out of which this bubble is blown, the 'soap-film', is empty space welded onto empty time"! Were there no emptiness, nothing, not even movement, could be conceived of. We could not even have the illusion of the City of the Gandharvas. "Because he is not full therefore when he becomes decayed he can renew". (Lao-tse). The whole question of the origins of the universe is extremely controversial at the moment, (it probably always will be), so I do not feel competent to discuss it here, I only wanted my reader to feel that emptiness is essential, whatever be the true nature of movement.

[8] *Men, Galaxies and Quasars.*

Emptiness of its very nature beggars description; one can only judge its existence by what it is not, and from hints gathered at various stages of the way. All I am trying to do is to give my reader the sense of it that has grown up in me, spontaneously, in the course of time. For many years I have been collecting references to it from various sources, from Buddhism to science, a word here, another there, and all bear out what Eckhart said six hundred years ago: "All things have been born out of nothing, therefore is the nothing their true origin".[9] Eckhart knew what modern science is now discovering once more. It is probably a facet of Reality?

Where then can we find unmistakable signs of the prevalence of this "nothing"? In any attempt to gauge what the universe really is, it confronts us at every step, only we turn away our eyes, perhaps because the idea is not particularly pleasant? But as we have set out in search of "reality" wherever it is to be found, we may as well consider what the sciences of physics and astronomy can tell us about it; after all it is their business to find out.

Apart from the "creation out of nothing", Peierls tells us that: "Rutherford concluded that the atom was mostly empty and that all heavy parts, (heavy comparatively speaking LCB) which it contained, as well as the alpha particles he was using as his cue ball, must be very small compared to the size of the atom",[10] which would mean that the rest is emptiness. The same applies to galaxies between the arms of which the same emptiness can be found. John Pfeiffer gives us some

[9] Buttner transcription, vol. I, p.101. [10] Opus cit., p.96.

idea of this situation: "Strictly speaking there is no such thing as 'pure space', a perfect vacuum. Utter emptiness is unknown in the *observable* universe. But nature comes mighty close to it. There are regions where matter (or motion LCB) is sprinkled so sparingly that nothing, absolutely nothing, in our experience can serve as a basis of comparison". He himself comes nearest to it: "A gnat buzzing alone in the barren skies of an uninhabited earth, or in a volume of space as large as that occupied by the entire earth, would be less alone than an atom wandering through the spaces between the galaxies"![11] Where no movement streaks across such spaces, there is nothing!

But these are not the only circumstances where such voids are to be found. Actually they would seem to be at the very fountain-head of all we know: at the entrance to the brain. As Eddington said: "The physiologist can trace the nerve mechanism up to the brain; but ultimately there is a hiatus which no one professes to fill up. Symbolically we may follow the influences of the physical world up to the door of the mind; they ring the door-bell and depart".[12] Supposing on a vast scale or an infinitesimal one, motion stops, then goes on, we would obtain the same effect as those black spots on the photographic plates of the electron, but instead of a particle-wave, it would be "nothing between", of which I have spoken before. (See "Neti Neti"). In photographs of the Trifid Nebula in Sagittarius, tiny black spots which they call "globules" have been discovered, but they look like holes in the ocean of light. Astronomers think these may be stars on

[11] Opus cit., p.153. (italics mine). [12] *Nature of the Physical World*, p.89.

the verge of being born—out of emptiness! If that be so, may it not be possible to look at the electron from the same point of view? Supposing the whole theory of electron as both wave *and* particle were to be regarded from a new angle? This is not at all impossible seeing that the "particle" aspect has never actually been detected by any manner of means. What if there were no such particle but only gaps or holes out of which the electron-*wave* draws fresh momentum? It is surely a possibility? Wherever there is motion, there those black spots are to be found, in our bodies as well. Dr. Emmet says: "There is a gap in which a transmutation occurs of which no satisfactory account has yet been given, between agitations in the cortex and conscious sensation".[13] And what is extremely interesting in this connection is that such transmutations are not only to be found in human physiology, but even in the history of species, occurring over lapses of years or centuries. Here also in entirely different dimensions, there are the same gaps or plunges into emptiness which are so unaccountable.

Erwin Schroedinger tells how the Dutchman, De Vries, "discovered that in the offspring even of thoroughly pure bred stock, a very small number of individuals, say two or three in tens of thousands, turn up with small but 'jump-like' changes, the term 'jump-like' not meaning that the change is so very considerable, but that there is a discontinuity, inasmuch as there are no intermediate forms between the unchanged and the few changed. De Vries called that a mutation. The significant fact is the discontinuity.

13 *Nature of Metaphysical Thinking*, p.58.

(What seems to me significant in all this is that *out of that discontinuity arises a new form of life*! LCB). It reminds a physicist of the quantum theory—no intermediate energies occurring between two neighbouring energy levels . . . Mutations are inherited as perfectly as the original unchanged characters were".[14] These mutations are rare events whilst the electron makes the quantum jump in fractions of a second, but both seem to be following the unalterable necessity of a sudden lapse into emptiness; and I can only repeat: what is time when one is following universal laws? As far as we can tell these "gaps" are perhaps resting places, emptiness of motion, a taking up of new energies? Or is it in all cases a plunge into an emptiness out of which new birth will arise?

We have been considering these gaps in the process of our nervous systems, of races, of atoms and of galaxies, as manifestations of movement and emptiness. But may there not be something more? Eckhart seems to have surmised something more—as also did Lao-tse. "For its (the light's) ground is a uniform stillness which in itself is motionless, and out of its immovableness all things are moved and all receive life who live according to Mind".[15]—that "Mind" which the Lankavatara Sutra says is the only thing there is. But how should we live according to it?

[14] *What is Life?* p.34.

[15] Franz Pfeiffer transcription, p.194, (author's translation from the German).

II

THE ANSWER TO THAT QUESTION LIES behind all I have been saying so far for it raises the further question: what is mind? Eddington said that we should never forget that: "Mind is the first and most direct thing in our experience; all else is remote inference".[1] Of such inferences, the most important one for us, it seems to me, is that we should "infer" that we are parts of the same motions as those of neutrinos and stars—merely in another dimension.

Movement and emptiness—we might also regard them from the point of view of life and death, for, as I see it, death "emptiness" is not annihilation: "Hollow it seems, nothing is lacking, if it is moved more will it bring", as Lao-tse said. We saw that the smallest motion has repercussions throughout the universe, therefore although we admit that these motions take place across gulfs of emptiness, yet seeing that they always catch up again on the other bank, we may accept it that there is more in this "hollow" than we had imagined. If we accept death as the same "hollow", we may perhaps conceive that out of it,

[1] *Science and the Unseen World*, p.24.

"more" will be "brought". Be that as it may, I propose to look at deep sleep and death (which to our waking consciousness present much the same appearance—namely a blank) from the point of view of those "jumps" we have been speaking of,—Eddington's "hiatus",—remembering that time as such does not make either life or death more of a reality. Undoubtedly this is a question that interests us as deeply as any of the other aspects of movement and emptiness of which I have spoken thus far.

Considering first of all the question of dreams, round which scientific experiments are now in progress. These have thrown an entirely new light on the nature of sleep and shown it to possess qualities unsuspected hitherto. By means of the electro-encephalograph (EEG) it has been discovered that weak electrical currents are continuously generated through the skull by millions of pulsating cells. This brain electricity was first discovered by the German neurologist Dr. Hans Berger in the 1920s. "Tiny electrodes pasted to the scalp carried brain waves out at from 8 to 13 cycles per second, making a characteristic rippling signature. These were called Alpha rhythms. In deep sleep . . . the waves are large and slow and have a frequency of from $\frac{1}{2}$ to 2 per second—the Delta rhythms".[2]

Later it was decided to monitor eye movements during sleep by pasting extra electrodes around the eye. To everyone's surprise it was found that there were not only rolling eye movements but also fast outbursts of

[2] Edwin Diamond, *The Sleeping Mind*, p.85, et. seq. All following references are to this, unless otherwise stated.

eye activity. "Whereas the individual eye movements took less than a second, the clusters might last anything from 3 to 55 minutes". Going back over the EEG records, Kleitman in the USA noted that there were changes in the brain-wave pattern associated with these bursts of rapid movements (REMs): "Apparently something emotional and engrossing was going on in the sleeping mind . . . Kleitman had the brilliant hunch—(he said: 'Call it a hypothesis')—that the sleepers were dreaming, and the test which followed, showed indubitably that he was right." Diamond says: "The historic results were reported in the journal 'Science' of September 4th 1953 . . . A whole New World lay ahead".

I will not go into details here: anyone interested enough can study this and subsequent works on the subject, but I must give the most important results of William Dement's experiments with the tape-recorder that gave 2-300 yards of brainwave patterns. Dement found that: "all the REM bursts always made their appearance when a distinctive fast 10-per-second brain-wave pattern was being traced on the graph paper. This pattern resembles the Alpha pattern of wakefulness . . . Moreover when the fast pattern ended, the REMs ended . . . (See fig. 1). This nightly pattern is as universal as sleep and as regular as the motions of planetary bodies",—we always come back to cosmic conditions! But seeing that this pattern holds something even more significant for me than are sleep and dream, I have to consider it from various points of view.

Later experiments have shown that every one (even those who declare they never dream) dreams in regular

Figure 1

deep sleep – valley pattern traced by brain's activity indicates all quiet

Dream emerged. Distinctive brain-wave clusters signal emergence of dream

The Dream Pen records flat pattern characteristic of dream thought

The Dream goes on. Eye electrodes record how the sleeper "watches" the show

Figure 2

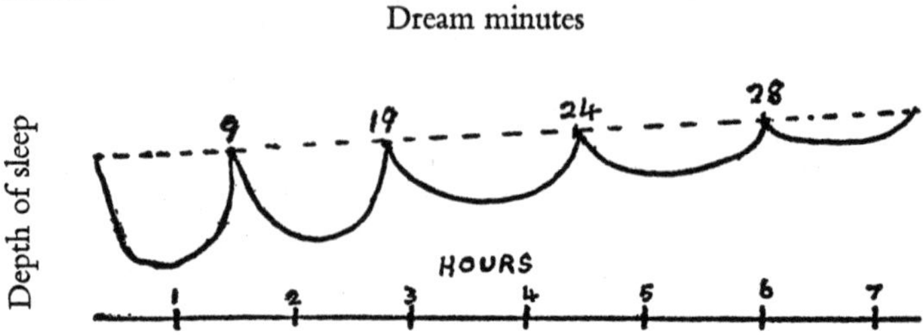

Reproduced by kind permission of Edwin Diamond

cycles alternating with dreamless sleep, every night. The EEG indicates by different patterns of electrical activity that during a typical night of 7 hours of healthy sleep, we spend more than $1\frac{1}{2}$ hours, or twenty per cent of the night in dreaming. A study of Fig. 2 shows that in the beginning there is profound unconsciousness in deep sleep,—what the Upanishads call "prajna". Although the sleeper is drawn down into emptiness between periods of mental activity in dream, yet each time the depth decreases whilst the periods of dream activity become progressively longer; from 9 minutes at first to 28 minutes the last time; finally both fuse in the awakening mind. What is interesting is that as the gaps become more shallow and the periods of motion longer, these increase at diminishing ratio. If we believe in reincarnation, is it not conceivable that lives become progressively deeper (corresponding to the longer periods of dream) but becoming more and more difficult at diminishing ratio as we evolve, until the final culmination in the Mind-only, foreshadowed by the Lankavatara Sutra?

This science of dreams is still too much in its infancy to have tackled the problem of the content of dreams, and dream "scientists" have not even begun to turn to C. G. Jung for his experience in this matter, which was vast. (Jung before his death had become deeply interested in the work being done on sleep and dreams in America). But whilst Jung's data for the *substance* of dreams has not yet been taken into account, we have to content ourselves with physical discoveries. These have brought two most interesting facts to light: firstly that our brains work at regular intervals of time, between which there is a pause, an apparent blank. But do we not find this same blank in different conditions throughout the body? The heart 'rests' between beats, for longer times in some people, shorter in others; some have 80 beats a minute, some 60. The conscious life of man is governed by messages from the senses to the brain, but as we learnt from Eddington, there is a hiatus between the influences from the outer world and the form in which they reappear in consciousness on the other side of the gap. In sleep that hiatus is 90 minutes, in other cases a few seconds, neurologically some hardly reckonable portion of a second—yet all the same everywhere there is a pause in whatever movement it be, between one period of activity and another.

The second thing that this work on dreams has made clear is that we could not lead our lives without both dream and sleep. The longest time anyone has been able to exist without sleep and remain sane is 201 hours 13 minutes. It seems that a continued and complete loss of sleep will eventually precipitate insanity and death. (*Tortura insomniae* was used in mediaeval times to put witches to death and I am not sure that it has not been used in modern torture chambers!) But what is an entirely new discovery is that dreaming is as necessary to us as is sleep itself. Robert Wilkinson says that: "although there were not any clear signs of neurosis consequent on people being woken up from their dreams, what did happen was of great importance for the development of our knowledge of sleep. The subjects tried to catch up on their dreaming sleep. They had to be woken up progressively more often and when allowed to sleep undisturbed, they spent much more time than usual in this dreaming state of sleep . . . The conclusion must be that this dreaming sleep is a unique kind of sleep, different in nature from that in which no eye movement occurs . . . *We need both sorts and one is no substitute for the other*".[3] That would seem to apply to all species of animals: who has not watched a dog or cat dreaming on the hearth? It is probably only the periods that differ.

At present it is impossible to know the cause or effect of these two forms of sleep in relation to our mental state, the only thing we can be sure of, is, that there is a close relationship between the two. Obviously the eye movements show that in sleep we are seeing things, but what? Of course directly our brain waves are applied to them, we form them into some kind of image familiar to us as the City of the Gandharvas. Jung goes much further than the physical aspect of dreams, he puts them in their psychological setting and studies them in relation to the life of the dreamer. I do not speak of "memory" dreams, which of course have always been recognised, but of prophetic dreams which Dunne also had—equally this author. Jung collected signs of such dreams which provide unmistakable evidence of design. Such design would seem to show another kind of condition behind sleep and dream—is that perhaps the meaning of emptiness?

Regarding this matter of sleep and dreams: is it not possible that dreams contain the raw material of our famous "City"? The brain cells may have free play in sleep and throw out image-patterns in waves which,

[3] See: Sleep and Dreams, *Listener*, 14/4/66. (italics mine).

in sleep are usually random collections of dream appearances, but later on waking, become co-ordinated into the images of our daily lives. I think we might find grounds for this conjecture in the various forms of brain-waves recorded during sleep. Apparently: "The brain-waves of dreaming sleep are similar to those *present during wakefulness.* They are of low amplitude, rapid and of no clear frequency. The normal sleep on the other hand is associated mainly with synchronous waves of more definite frequency and in the deeper stage of this sleep, slow waves of large amplitude predominate at a frequency of about 1 to 3 per second"[4] Seen from the waking point of view, J. Z. Young writes of these brain-wave frequences: "Seeing that the activity of nervous conductions (in the brain) is accompanied by electrical changes . . . there is no doubt that the electrical brain waves are in some way a sign of the activities of the brain cells . . . (As Radio waves are the signs of star 'cells' LCB). It may be significant that the waves are most clear when the brain is idling, —for instance when a person is asleep or day-dreaming when apparently large masses of cells are working in unison. When a person wakes or begins to think, the electric changes become more complicated".[5] Which confirms what Wilkinson said of the sleeping and dreaming brain!

Of course so far, no one can claim to know the reason for these mysterious differences in wave-lengths under different conditions, but everyone has the right to draw his own conclusions, and I would like to give mine, to which all I have been saying about

[4] Wilkinson, opus cit. [5] *Doubt and Certainty in Science,* p.56.

movement and emptiness have led me. And, being a believer in re-incarnation, what I say concerning the nature of sleep and dream waves, applies equally to the waves of life and death, to my mind—what in the East is called "Karma". Having understood the activity of atoms, we can no more regard death as final.

I should like to turn back for a minute to what was said of the electron: we may remember that: "the spreading of waves in space did not mean that the electron spread; it meant that the probability of our finding the electron spreads as the wave does". Therefore there is clearly a gap between where the electron is at one moment, (or where we suppose it to be) and where we next discover it. That is its wave nature, which carries throughout the universe. Supposing we regarded brain-waves from that point of view, and, reverting to Fig 1, imagine the low level points in the deep sleep pattern to be those gaps between one appearance of brain-wave and the next? In deep sleep, as we saw, we do plunge into emptiness, the source from which our lives gain strength. As this is fundamental (compare the pre-natal stage), it goes more deeply down and is more complete unconsciousness than any other hiatus. Then that time spent in emptiness gets more and more shallow as the effects formulate themselves in dream and ruffle the surface, till finally depths and surface merge in an awakening. My reader may here object that, although Fig. 2 shows the gaps into emptiness between periods of activity, the dream chart shows nothing of the kind. But we should remember that those dream vibrations take place with the rapidity of all pulsating brain motions. The EEG's recording instruments are probably not fine enough to register the gap at the base of a dip. They can register the gaps in sleep, these being in periods of over an hour, but it is a different thing when they take place in the nth of a second. We may remember what takes place at the "door of the mind"! I think that from all the information which we have collected regarding the nature of waves in the universe, we may be quite sure that these follow the same pattern throughout—the electron being prototype of all.

The brain-waves of man in deep meditation have lately been taken by EEG,[6] but though this recording may throw some light on what happens to men *physiologically* in that state, it cannot explain the experience of the Saint or Yogi who achieves such "liberation in life". And Leggett admits that it is not likely that research on physiological lines can ever achieve that. Therefore we may accept the possibility that those who attain what is called "enlightenment"—be it Saints, Bodhisattavas, Tulkus or whatever name any religion may give to them—have reached the state of being able voluntarily, to sink themselves in the "emptiness", from whence all life. "Therefore should the spirit step out over objects and all objectivation, over forms and all formulations, even over being and its differentiations, then will dawn in it the complete reality of bliss which pertains to the creative mind alone."[7]

When I started on the study of dream conditions, I thought that the difference between the ascending

[6] See Trevor Leggett: *The Tiger's Cave*, p.185.
[7] Eckhart, Buttner trans., vol. I, p.203.

road from deep sleep to dreaming, from death to life, was that, in the one case one could trace the way of this transformation; in the other, one could not, the origins being beyond our ken. Now however I realise that in both cases we must forever ignore the sources, for they lie in emptiness. What is it that manifests in the dreaming and waking stages and becomes a blank in the empty stages? Is it perhaps ripples of mind across unmoved depths? The Lankavatara says: "Analysed down to atoms there is indeed no form to be discriminated as such; what can be established is the truth of Mind-only".[8] Considering all the groping after Mind that has gone on for thousands of years, can we say that we are any nearer to knowing what it is to-day than were the authors of the Lankavatara? Sherrington knew the truth of this: "Mind for anything perception can compass, goes therefore in our spatial world more ghostly than a ghost. Invisible, intangible, it is a thing not even in outline; it is not a 'thing'. It remains without sensual confirmation and remains without it forever. All that counts in life".[9] One cannot say more.

[8] Ch. XIII. [9] Opus cit., p.256.

III

SHOULD WE THEN ASK OURSELVES: IS AN awareness of this Mind movement the highest stage we can hope to arrive at or may there be some yet higher form not even guessed at? Of course if we compare our own to other aspects of Mind on earth, ours is undoubtedly the most evolved. But then, if he had self-consciousness, probably the elephant would think the same; and who is to say that there are not other forms as much more evolved than ourselves, as we than the elephant? We must realise that this oft-quoted "City of the Gandharvas" is that which the human brain cells, plus senses, have devised on this planet Earth, but who can tell what agglomerations of movements and emptiness may look like to the inhabitants of other planets in our or other galaxies? Their brain cells (if they have any) probably create something utterly different, out of the same nothingness. Eckhart said: "God works in the non-existence". For the Lankavatara also, Mind-only is Sunyata—emptiness. Both knew the same truth, but for ordinary man it still remains most difficult to understand—or accept.

If we are prepared to admit the truth of all I have

been saying, the question must arise: how are we to apply it to our common life, which is after all, for us the most important question. It seems to me that, to begin with, we should never let our minds get fixed—fixed on an object, fixed in dogma, of whatever kind, religious or political. Once it gets fixed on one point, there is a stoppage and we are no more flowing with the movement of the universe. According to Takuan, the greatest Japanese exponent of this idea: "The thing is not to try and localize the mind anywhere but to let it fill up the whole body, let it flow throughout the totality of your being. Instead of localizing, or keeping it in captivity, or freezing the mind, let it go all by itself freely, unhindered and uninhibited. It is only when this is done that the mind is ready to move as it is needed all over the body with no stoppage anywhere".[1] The Japanese illustrate this idea with examples from the arts of archery and swordsmanship. In using the sword for instance, "do not think of how to kill the enemy, do not think how to protect yourself, or how best to use the sword", in fact any fixed intention stops the mind at one point, neglecting the rest, thus inhibiting the freedom of action and making you vulnerable.

But I do not need to go as far as the Samurai lore to illustrate my meaning: rightly understood we could all practice this non-fixation, what Zen Buddhists call "letting go". As an example from a recent experience: a short time ago I had to go to a place in the south of England, about 40 miles out of London mostly through urban areas. I do not know South London at all, nor

[1] D. T. Suzuki: *Zen and Japanese Culture*, p.107.

the country to which I was going. I was told it would take me $1\frac{1}{2}$ hours to reach my destination. I started off without a map, and only knowing the general direction in which I was heading. A chauffeur told me: take the A30 road, through Streatham and go by the Purley Way. How could I know where that "Way" was, or the roads through Streatham which I should take? Yet I struck Purley Way at its precise junction with the main street, and later took all the right turnings without hesitation, and, following roads 23 and 22 instead of 30, reached my destination $\frac{1}{4}$ of an hour sooner than expected. Going back I fixed my mind on finding the way by which I had come, carefully observing road-signs, and taking main streets—I lost my way three times! On my way out I could not have fixed my mind as I had no knowledge of the way to my goal any more than the Zen archer can see his target in a mist. He shoots his arrow into the Bull's eye, *because he does not use his reason but allows his mind to use him*, acting by what the Japanese call "mushin-no-shin", mind of no mind. If we could achieve that, we would be in harmony with the movement of the universe and could not go wrong. But this is usually screened by our intellects.

That is perhaps what the Diamond Sutra means when it says: "Awaken the mind without fixing it anywhere". Our minds should be continuously receptive, flowing through the moments that pass, picking up "what is" by the road-side. Everywhere the universal movement through emptiness has its particular quality to contribute to the whole, but a fixation blocks the stream and causes us to fail.

Dr. Suzuki said: "When it (the mind) is nowhere, it is everywhere".[2] That experience of mine is only a very inadequate example of what might happen to us if we could achieve that. And surely the first step would be to recognise our path—not to plan *the way* to go, but to allow something else to take over, something which knows, and to which the map of life is as clear as is a page of an atlas before our eyes,—as it took over that day when I drove south. The importance or not of the end to be achieved is of no consequence, the only thing that matters is that mind should be allowed to act unhindered in small as in great.

But before reaching such complete abandon to the nature of the universe, we have a very long way to go, only the rarest have achieved it in human history. The authors of the I Ching knew, and gave us a means by which we might learn the first steps, because only by letting go can we interpret its message. People like Jesus and the Buddha reached the final stage possible to man thus far, in the transference of the body to Mind-only. And I think that no one will deny that they were like that fly of which Pfeiffer spoke but their wings got caught, not in the intergalactic web but in that of this Earth which has trembled ever since. Have they not radiated throughout mankind for over 2,000 years?

Perhaps if we avoid destroying ourselves in nuclear warfare, and stave off starvation long enough—if computers take over our brain-work and our intellects atrophy sufficiently, the universal Mind will emerge from its swaddling clothes of physical apparitions and

[2] Opus cit., p.108.

reign triumphant in a human race conscious of its true nature where heretofore it had been unconscious.

For of course there is only ONE ALL containing everything—the furthest galaxy, the nearest atom, the greatest and the smallest, movement and emptiness, past, present and future—in it they are and they are not, for one is all and all is one for ever and ever—a billion years in thy sight are but as one day. Only there is no "sight", IT IS.

"That which is one is one; and that which is not one is also one".